Suren Machiraju and Suraj Gaurav

Power BI Data Analysis and Visualization

Suren Machiraju and
Suraj Gaurav

Power BI Data Analysis and Visualization

—

DE
G
PRESS

ISBN 978-1-5474-1678-3
e-ISBN (PDF) 978-1-5474-0072-0
e-ISBN (EPUB) 978-1-5474-0074-4

Library of Congress Control Number: 2018949269

Bibliographic information published by the Deutsche Nationalbibliothek
The Deutsche Nationalbibliothek lists this publication in the Deutsche Nationalbibliografie;
detailed bibliographic data are available on the Internet at http://dnb.dnb.de.

© 2018 Suren Machiraju and Suraj Gaurav
Published by Walter de Gruyter Inc., Boston/Berlin
Printing and binding: CPI books GmbH, Leck
Typesetting: MacPS, LLC, Carmel

www.degruyter.com

With a deep sense of gratitude, I dedicate this book to my parents-in-law—Mrs. Bhanu Bollapragada and Dr. B.K.B. Rao. Thank you for all the love and support.

—Surendra Machiraju

This book is dedicated to my parents and in-laws for their constant support and inspiration.

—Suraj Gaurav

DOI 10.1515/9781547400720-201

About De|G PRESS

Five Stars as a Rule

De|G PRESS, the startup born out of one of the world's most venerable publishers, De Gruyter, promises to bring you an unbiased, valuable, and meticulously edited work on important topics in the fields of business, information technology, computing, engineering, and mathematics. By selecting the finest authors to present, without bias, information necessary for their chosen topic *for professionals*, in the depth you would hope for, we wish to satisfy your needs and earn our five-star ranking.

In keeping with these principles, the books you read from De|G PRESS will be practical, efficient and, if we have done our job right, yield many returns on their price.

We invite businesses to order our books in bulk in print or electronic form as a best solution to meeting the learning needs of your organization, or parts of your organization, in a most cost-effective manner.

There is no better way to learn about a subject in depth than from a book that is efficient, clear, well organized, and information rich. A great book can provide life-changing knowledge. We hope that with De|G PRESS books you will find that to be the case.

DOI 10.1515/9781547400720-202

Acknowledgments

We wish to acknowledge the contribution of Rahul Jain and Jennifer Curiak. Rahul is an amazing technical editor and Jennifer does an amazing job of technical proofreading and cleaning the book up.

Thank you!

Suren Machiraju
Suraj Gaurav

DOI 10.1515/9781547400720-203

About the Authors

Suren Machiraju developed an innovative supply chain solution integrating online stores with market makers and aggregators, founding Commercia Corporation in the late 1990s. Within one year, Microsoft acquired Commercia Corp, providing Suren the opportunity to lead the business-to-business (B2B) interoperability team within the BizTalk business unit. Over the next six years, Machiraju's team delivered five releases of the BizTalk Server (2000–2006R2). Subsequently, Machiraju led the BizTalk Rangers-Customer Advisory Group, and in two years, lit up over twenty of the largest middleware deployments on the .NET stack.

In 2011, Suren collaborated to create the Azure Customer Advisory Team at Microsoft. For five years, Machiraju led efforts in engaging enterprise customers, start-ups, and partners for architectural reviews and deployments of cloud/hybrid cloud .NET and OSS applications on the Azure platform. The team pioneered solutions for the most challenging cloud projects producing dozens of successful deployments.

In 2014, Suren accepted an appointment as a Technology Business Partner at the Bill & Melinda Gates Foundation where he collaborates with leading nongovernmental organizations and nonprofit partners in devising technical solutions for some of the world's most challenging social issues.

Machiraju holds a master's degree in Mechanical Engineering from the Birla Institute of Technology and Science in Pilani, India. He is a listed author of over twenty patents in business software areas of B2B and Data Interchange Standards and has published books and authored dozens of MSDN articles/technical blogs on Azure and .NET. When he is not publishing blogs or presenting works to the larger technical community, he is enjoying time with his family in the beautiful Pacific Northwest and cheering on the Seahawks each Sunday.

"Please contact me if I can be of assistance in architecting your cloud-based solution, collaborating in this space is one of my greatest passions"—Suren. https://about.me/surenmachiraju

DOI 10.1515/9781547400720-204

Suraj Gaurav started his career in 2000, at the height of the dot-com era. He worked in a start-up, Asera, that was developing a revolutionary platform for building business-to-business applications. In 2002, he moved to Seattle to work for Microsoft. He spent almost ten years there and worked on various products including BizTalk Server, the Commerce platform, and Office 365. He has in-depth experience building enterprise-scale systems like BizTalk, to internet scale services like Office 365. He also built the consumption-based billing platform serving as the commerce engine for Azure.

Gaurav holds a bachelor's degree in Computer Science from the Indian Institute of Technology, Kanpur, India. He is listed as an inventor in over twenty-five patents. When he is not working, he can be found spending time with family and enjoying the beautiful outdoor life of the Pacific Northwest.

About the Technical Reviewer

Jennifer Curiak specializes in Dynamics 365 implementations, agile coaching, project management, business analysis, quality assurance, and technical writing. She works to help teams in a variety of industries become more productive, communicate more effectively, and generally get stuff done.

A writer at heart, Curiak started her career as a technical writer for a software company in 2000 and has evolved into designing solutions, managing QA processes and resources, coaching large and small teams in agile development practices, acting as scrum master, and working on Dynamics 365 customizations and implementations. She was the technical reviewer on the book *Administering, Configuring, and Maintaining Microsoft Dynamics 365 in the Cloud in 2018*, continues to write in-house technical documentation and end-user documentation, and contributes to other professional publications.

Jennifer and her husband Mike live in Western Colorado and spend most of their free time exploring empty and desolate areas of the West by mountain bike and packraft. She can be contacted directly at jcuriak@inotekgroup.com.

Jennifer Curiak
C) 720-933-5587

DOI 10.1515/9781547400720-205

Foreword

It is with great pleasure that I write this foreword for *Power BI Data Analysis and Visualization* authored by my friends—Suraj and Suren.

Data Analysis and Visualization is a vast horizon, especially in a world where over 2,500,000,000 GB of digital data is created every day. With the advent of IoT and automation that number is expected to grow exponentially. Leveraging those immense amounts of data through Data Analysis and Visualization is a key component to profitability.

In this book, Suraj and Suren focus on Microsoft's Power BI and clearly demonstrate to the readers via precise instructions how to build data visuals on a variety of database and CRM applications. As a bonus, the reader also gets a tutorial on embedding the visuals in an Azure Application. Integration of the Cortana suite is indeed a multiplier.

My very best wishes to the authors and the community of readers.

John R. Girgis
PMP-Certified Project & Engineering Manager
https://www.linkedin.com/in/john-r-girgis-pmp-026a1720/

DOI 10.1515/9781547400720-206

Contents

DOI 10.1515/9781547400720-207

Introduction

Broadly speaking, data visualization is a term used to describe a way of presenting raw or tabular data in a more easily understood visual context. With data visualization, users trying to understand raw data can see relationships, trends, and patterns in a more clear and easy to understand manner. As such, data visualization can be used as an effective and efficient form of communication of the state of a business for those who need and want to see deeper into the information generated and captured by their business processes.

In its 2017 Report, Gartner estimates that *by 2020, organizations that offer users access to a curated catalog of internal and external data will realize twice the business value from analytics investments than those that do not.*

The most significant challenge for data visualization is accessing data. The most critical business data is locked up in silos within ERP and Custom Applications. Who are the vendors that provide state of art Data Visualization Tools? How does a Data Visualization Developer extract the data? This book provides you the answers to these questions.

Who Should Read this Book?

1. **Business Owners and IT Pros:** By 2021, the number of users of modern business intelligence (BI) and analytics platforms that are differentiated by smart data discovery capabilities will grow at twice the rate of those that are not and will deliver twice the business value.
2. **Data Scientists:** Through 2020, the number of citizen data scientists will grow five times faster than the number of data scientists.
3. **Data Analysts:** By 2020, natural-language generation and artificial intelligence will be a standard feature of 90 percent of modern BI platforms.
4. **Enterprise Developers:** By 2020, 50 percent of analytic queries will be generated using search, natural-language processing or voice, or will be auto-generated.
5. **Enterprise Architects:** By 2020, organizations that offer users access to a curated catalog of internal and external data will realize twice the business value from analytics investments than those that do not.

DOI 10.1515/9781547400720-208

What Will You Learn?

- Market survey of Data Visualization solutions
- How to use the common and advanced Power BI features
- How to deploy the embedded Power BI Dashboard as an Azure Application
- How to build a modern Power BI solution with Microsoft SQL Server and leverage Microsoft Stack including Cortana
- How to build a visually gratifying Power BI solution on open source data stores, that is, PostgreSQL
- How to unlock enterprise secrets, such as spotlighting the most relevant business trends by integrating Power BI with Dynamics CRM and using Natural Language Queries to tease out trends.

We appreciate your investment in this book. We would love to hear from you so that we may improve this and future offerings.

Chapter 1
Introducing Data Visualization

Overview of Technology

Data visualization is the concept of presenting data through visuals, such as info-graphics, charts, Sparkline, and geographic maps, etc. The presentation of data through visuals makes it more appealing to users and allows decision makers to view data analytics in a visual or graphic format. It helps users to easily identify difficult concepts or new patterns related to data. Data visualization tools help to easily recognize patterns, trends, and correlations that are otherwise difficult to see in textual data. In addition to interactive visuals, you can drill down into charts and graphs to access the details of the information displayed. You can also process data as per your requirements.

Importance of Data Visualization

It is easier for the human brain to process the graphical visual representation of data versus textual information. For example, charts or graphs can represent large complex data in a more clear and effective manner than in a spreadsheet or textual report. Data visualization provides an efficient way to communicate concepts in general.

Data visualization helps in:
- Identifying areas of enhancement
- Identifying factors that influence customer behavior
- Defining a presentation place for the products
- Forecasting sales volumes
- Democratizing data and analytics
- Presenting data-driven insights to all resources within an organization
- Creating big data and advanced analytics projects
- Interpreting complex algorithms easier than numerical outputs

DOI 10.1515/9781547400720-001

Data Visualization Tools

There are several data visualization tools available in the market. Some of them are listed as follows:
– Microsoft Power BI (This book focuses on this tool.)
– Tableau
– Qlik

Understanding Power BI

Power BI is a business analytics reporting tool introduced by Microsoft to create interactive business reports. It incorporates several analytics features to provide business insights across an organization. Prior to Power BI, end-users were dependent on information technology staff and database administrators for creating business reports and dashboards. Now, end-users can create business reports and dashboards on their own with the help of Power BI. The Power BI Datasets feature allows users to represent data and create visualizations based on the data. The concept of Power BI is illustrated in Figure 1.1.

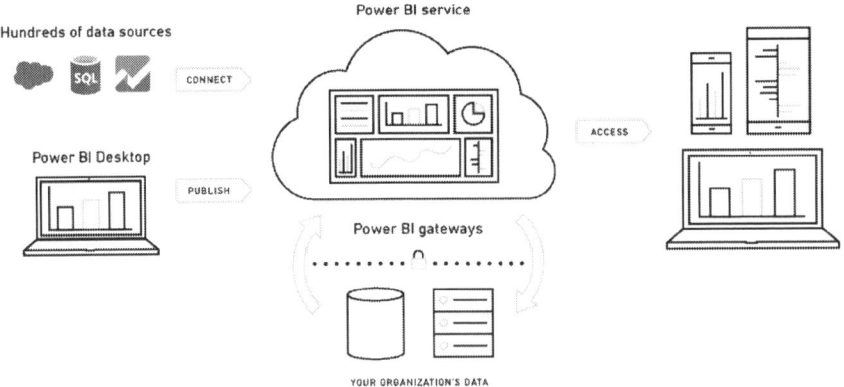

Figure 1.1: Concept of Power BI

Comparing Microsoft Power BI and Tableau

Tableau and Microsoft Power BI are listed as the top competitors in the list of the BI and data visualization tools. Both tools are easy to use and support a large collection of visuals to present data in a visualized format.

Table 1.1 lists the comparison between Microsoft Power BI and Tableau on several parameters:

Table 1.1: **Comparison between Microsoft Power BI and Tableau**

Parameter	Data Visualization Tools	
	Microsoft Power BI	**Tableau**
Overview	Provides a full overview of the most critical data of an organization.	Leverages data discovery for any type of business user.
User interface	Provides an easy-to-use interface like Excel.	Provides an easy-to-use interface.
Infrastructure support	Provides better support than Tableau.	Provides limited infrastructure support.
Dashboards	Provides scalable dashboards that allow users to select the desired visualization as a blueprint and use the navigation to drag and drop data into the visualization.	Allows users to embed dashboards into the available business applications, such as SharePoint, for quick data analysis.
Data sources	Supports several data sources such as SAP HANA, JSON, MySQL, etc. It examines the relationship between data sources automatically when data is added from multiple sources. It also allows users to connect to Microsoft Azure databases, third-party databases, files and online services such as Salesforce and Google Analytics.	Supports several data connectors including OLAP, NoSQL, Hadoop, and cloud options. It examines the relationship between different data sources automatically when data is added from multiple sources. It also allows users to apply changes to data links as per the requirements of the company.
Visualization support	Helps in uploading the datasets easily. It provides numerous visualizations that can be selected by users as blueprints. Data for the selected visualization can be inserted from a sidebar. It also allows users to use natural language queries for creating complex visualizations. To perform data analysis, it supports 3,500 data points.	Supports different types of visualizations, including heat maps and line charts, etc. It helps in easily creating complex visualizations through an interface without requiring coding expertise. It also allows users to use multiple data points in their data analysis.

Table 1.1 (continued)

Parameter	Data Visualization Tools	
	Microsoft Power BI	**Tableau**
Customer/technical support	Provides technical support for both types of users having either the free Power BI account or paid Power BI account. Both types of users can generate support tickets. It provides in-depth documentation to help novice and expert users through the step-by-step procedure of using Power BI. It also provides a common community forum for its users.	Provides several options for technical support. For example, a user can generate a support ticket directly through a phone call, or by sending an email to the customer portal. A user can also utilize the knowledge-based repository, as per the subscription category to which the software belongs. The community forum is also available for training purposes.
Pricing	Provides three subscription tiers including Desktop, Pro, and Premium. Power BI Desktop is free for individuals. Power BI Pro costs $9.99 per user per month. The pricing of Power BI Premium depends on the required capacity. Power BI Premium is charged a per node per month basis.	Is available in three tiers including Tableau Desktop, Tableau Server, and Tableau Online. There are two editions of Tableau Desktop including Personal Edition and Professional Edition. Tableau Desktop Personal Edition costs $35 per user per month. Tableau Desktop Professional Edition costs $70 per user per month. Tableau Server costs $35 per user per month. Tableau Online costs $42 per user per month.

Note
According to Gartner's report, "Microsoft Power BI is a leader in the Magic Quadrant for Business Intelligence and Analytics Platforms."

Key Features of Power BI

As discussed earlier, Power BI is a data visualization and business intelligence tool that allows business representatives to easily view the insights of their business in the form of interactive reports. Power BI provides several features. Some key features are as listed below:

– Free sign up
– Ability to receive data from multiple data sources
– Ability to obtain key metrics of your business
– Quick insights
– Data-driven decision making from anywhere

Sign Up for Free

One of the key features of any software is its cost. With Power BI, you can sign up for free. There is no need to save your credit card details. You can also easily monitor your data with the help of Power BI. You do not need any specialized setup or training to start your work. You can also utilize a trial Power BI Pro account to use the advanced features of Power BI. However, you will get only 10 GB space for the trial.

Receive Data from Multiple Data Sources

As discussed earlier, Power BI supports a large number of data sources including SQL Server, PostgreSQL, Dynamics CRM, and many more. You can transfer data into Power BI from these different data sources, analyze the collected data, and create interactive reports based on the gathered data.

Obtain Key Metrics of Your Business

Power BI provides a complete view of the key metrics of a business irrespective of the data source and the nature of data. You can get the complete view of your business data that might be coming from Excel spreadsheets, cloud services, or on-premises databases.

Quick Insights

The Quick Insights feature applies complex algorithms to a dataset in Power BI and quickly identifies different subsets of the dataset within the specified time frame.

Data-Driven Decision Making from Anywhere

As a business representative of an organization, you can manage your data from anywhere. You can use touch-enabled apps available for different devices including Windows, iOS, and Android. These apps can be used for accessing your organizational data.

Advanced Features of Power BI

In addition to the key features of Power BI, it also supports several advanced features. Some of the advanced features of Power BI are as follows:
– Ability to embed Power BI reports and dashboards into a Web App
– Real-time streaming
– Support for natural language query
– Ability to share content pack
– Ability to integrate with Cortana

Embed Power BI Reports and Dashboards into a Web App

One of the features of Power BI is the ability to embed your Power BI reports and dashboards into a Web app. This can be accomplished with the help of APIs and sample code available on a source repository like Git.

Real-Time Streaming

Real-time streaming is one of the advanced features of Power BI that helps users to stream data in Power BI dashboards in real time. In other words, the visuals pinned to Power BI dashboards are updated with the real-time data.

Support for Natural Language Query

One of the exciting features of Power BI is its support for natural language. You can query Power BI in your natural language (English) and get results in the form of visuals including charts and graphs. This feature is commonly termed as Q&A as you query from Power BI and Power BI provides answers to your queries.

Share Content Pack

Previously you were only able to share reports with other users in your organization. However, now you can share a complete package of your dashboard, report, and dataset with other users in your organization. This complete package is termed as a content pack. You can create the content pack and publish it to the team members. As you publish it, it becomes available in a centralized repository called AppSource. This repository helps the team members to easily locate reports and datasets published for them. All you need to create and access an organizational content pack is a Power BI Pro account.

Integration with Cortana

Now, you can integrate your Power BI reports with Cortana, an exciting feature of Windows 10 that quickly locates and lists the results to your queries made in natural language. The integration with Cortana provides relevant information directly from Power BI dashboards and reports. To use this feature, you need to create a Cortana answer page in Power BI, an Azure Active Directory (Azure AD)/Work or School account, and configure one or more datasets to be used with Cortana.

Variants of Power BI

As discussed previously, Power BI is a business analytics service that allows users to visualize and examine data quickly and efficiently. It provides user-friendly dashboards, interactive reports, and powerful visualizations through which users can be connected to a large number of data sources. There are two variants of Power BI, which are:
– Power BI Desktop
– Power BI Service

Power BI Desktop

Power BI Desktop, as the name indicates, is an on-premises version of Power BI. It allows users to build reports, queries, and data connections, etc. Users can easily share reports with others in the organization. Power BI Desktop embeds data modeling, visualizations, and Query engine. It integrates with Power BI Service, which makes data insights easier to build and share.

Power BI Desktop is a flexible and dominant tool that allows data analysts to:
- Connect to multiple data sources
- Structure data in an intuitive manner
- Create powerful data models
- Create coherent visuals

Power BI Desktop makes the process of designing and creating business intelligence reports simple and centralized.

Installing Power BI Desktop

As discussed in the previous section, Power BI Desktop is an on-premise solution. To utilize Power BI Desktop, users need to download and install it onto a local machine.

Perform the following steps to download and install Power BI Desktop:
1. Navigate to the following link:
   https://www.microsoft.com/en-us/download/details.aspx?id=45331
2. Click the **Download** button, as shown in Figure 1.2.

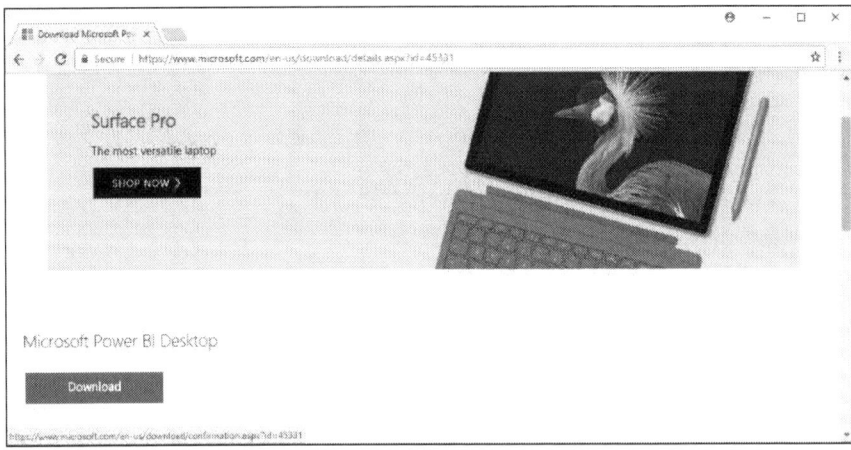

Figure 1.2: Downloading Power BI Desktop

The **Choose the download you want** page appears.

3. Select the checkbox beside the file name in the **File Name** column according to your system specification. If your system runs on the 32-bit operating system, select the checkbox next to the PBIDesktop.msi file. Otherwise, select the checkbox next to the PBIDesktop_x64.msi file.

4. Click the **Next** button, as shown in Figure 1.3.

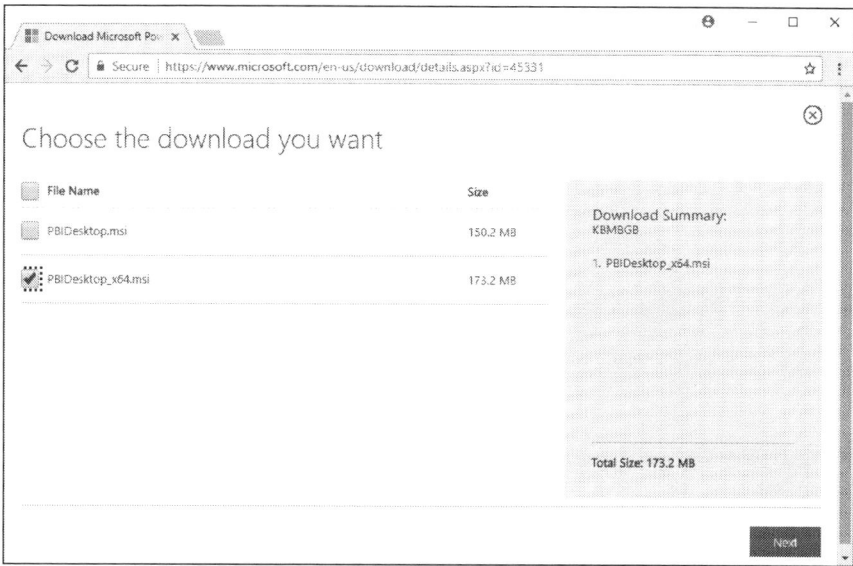

Figure 1.3: Selecting a file

The selected file is downloaded. You need to double-click the file to start the installation process. After double-clicking this file, the **Microsoft Power BI Desktop (x64) Setup** wizard appears with the **Welcome to the Microsoft Power BI Desktop (x64) Setup Wizard** page.

5. Click the **Next** button, as shown in Figure 1.4.

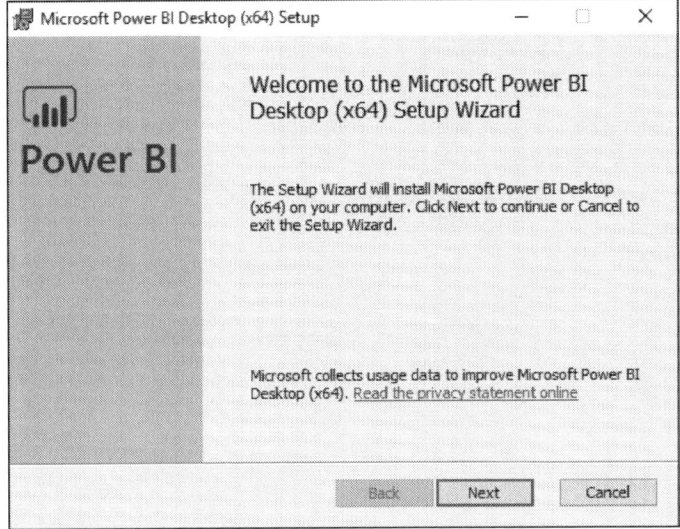

Figure 1.4: The Microsoft Power BI Desktop (x64) Setup wizard

The **Microsoft Software License Terms** page appears.

6. Read the license agreement.
7. Select the **I accept the terms in the License Agreement** checkbox.
8. Click the **Next** button to proceed to next page, as shown in Figure 1.5.

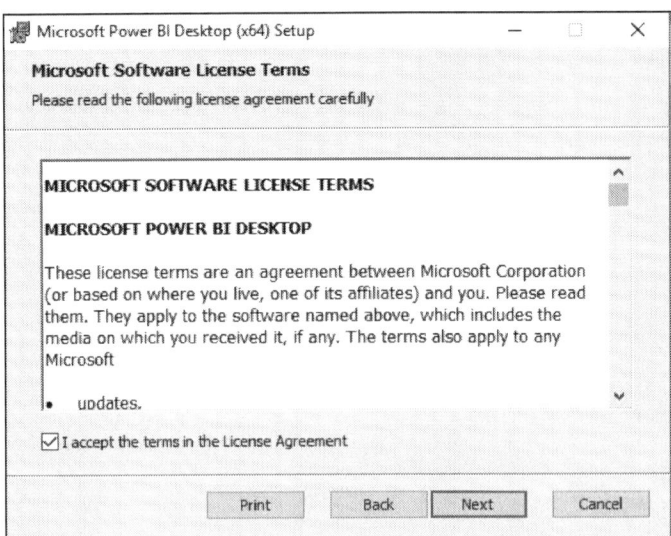

Figure 1.5: Accepting the license agreement

The **Destination Folder** page of the **Microsoft Power BI Desktop (x64) Setup** wizard appears.

9. Enter the full path of the destination folder in the **Install Microsoft Power BI Desktop (x64) to** text box.

10. Click the **Next** button, as shown in Figure 1.6.

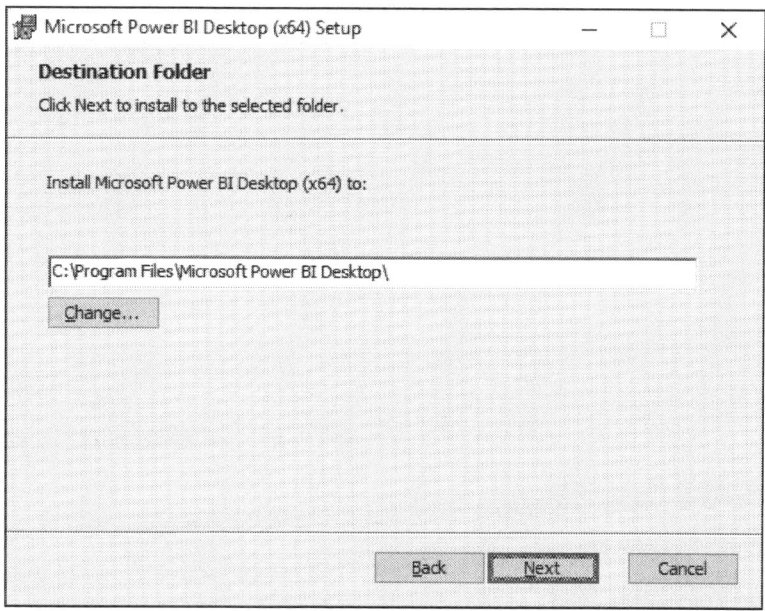

Figure 1.6: Specifying the destination folder

Note
You can click the Change button to change the path of the destination folder.

The **Ready to install Microsoft Power BI Desktop (x64)** page of the **Microsoft Power BI Desktop (x64) Setup** wizard appears.

11. Click the **Install** button, as shown in Figure 1.7.

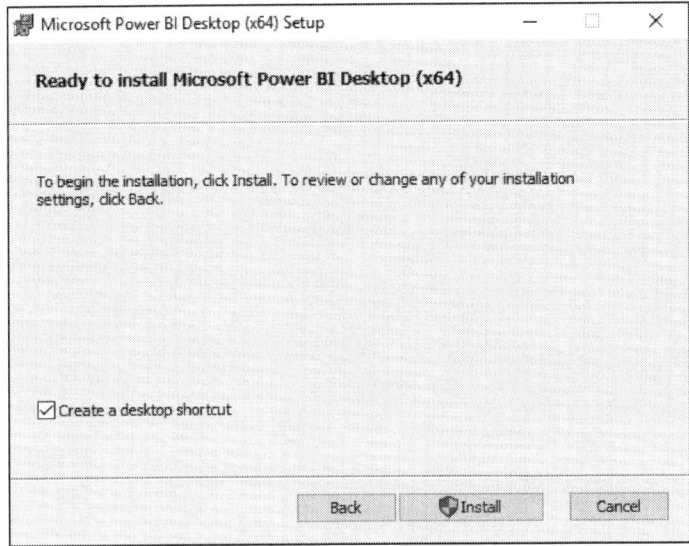

Figure 1.7: Installing Microsoft Power BI Desktop

The **Installing Microsoft Power BI Desktop (x64)** page of the **Microsoft Power BI Desktop (x64) Setup** wizard appears. This page displays the installation progress status, as shown in Figure 1.8.

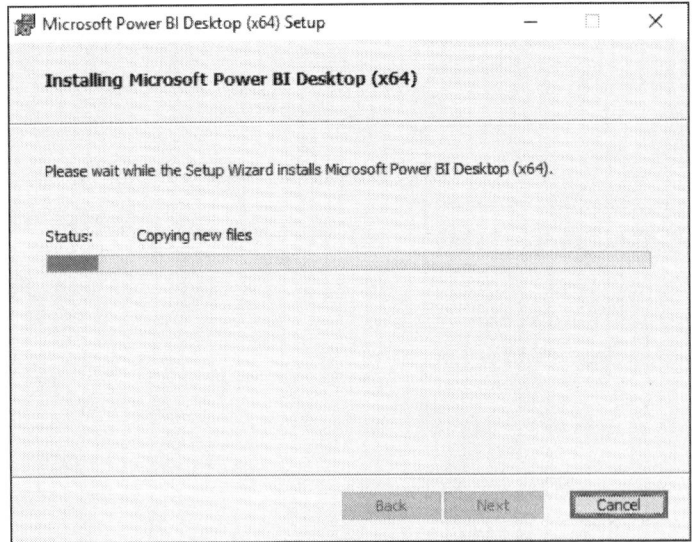

Figure 1.8: The Installing Microsoft Power BI Desktop (x64) page

The **Completed the Microsoft Power BI Desktop (x64) Setup Wizard** page of the **Microsoft Power BI Desktop (x64) Setup** wizard appears.

12. Select the **Launch Microsoft Power BI Desktop** checkbox to launch Power BI Desktop.

13. Click the **Finish** button to finish the wizard, as shown in Figure 1.9.

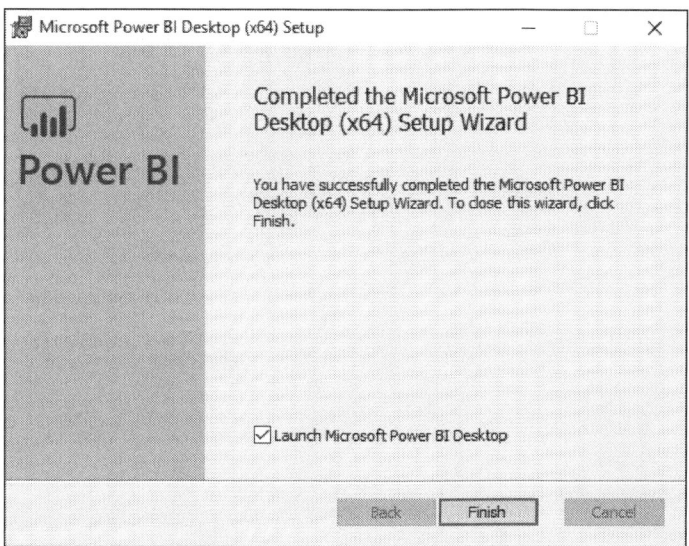

Figure 1.9: Finishing the wizard

The user interface of Power BI Desktop appears, as shown in Figure 1.10.

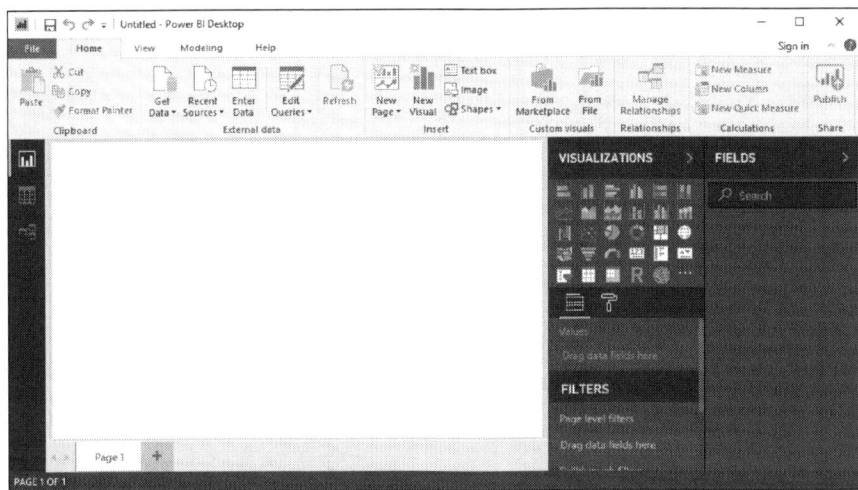

Figure 1.10: User interface of Power BI Desktop

Connect to Data

Before connecting to data, you should sign in to Power BI as it helps in collating Power BI Desktop and Power BI Service and allows them to work seamlessly. If you do not have the Power BI account, you can create it for free.

Perform the following steps to connect to data from Power BI Desktop:

1. Launch Power BI Desktop. The user interface of Power BI Desktop appears.
2. Click the **Sign in** link. The **Sign in** dialog box appears.
3. Enter the email address linked to Power BI in the **Email** text box.
4. Click the **Sign in** button.

The **Sign in to your account** dialog box appears.

5. Enter the relevant password in the **Enter Password** text box.
6. Click the **Sign in** button. Now, you are signed into Power BI Desktop.
7. Click the upper part of the **Get Data** button under the **External data** section of the **Home** tab, as shown in Figure 1.11.

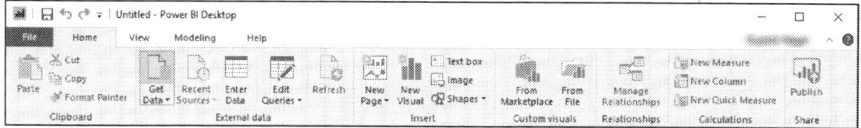

Figure 1.11: Clicking the Get Data button

The **Get Data** dialog box appears.

8. Select the **All** tab from the left pane. A list of available data sources appears in the right pane.

9. Select the desired data source from where you want to get data. In our case, we have selected **Excel**.

10. Click the **Connect** button, as shown in Figure 1.12.

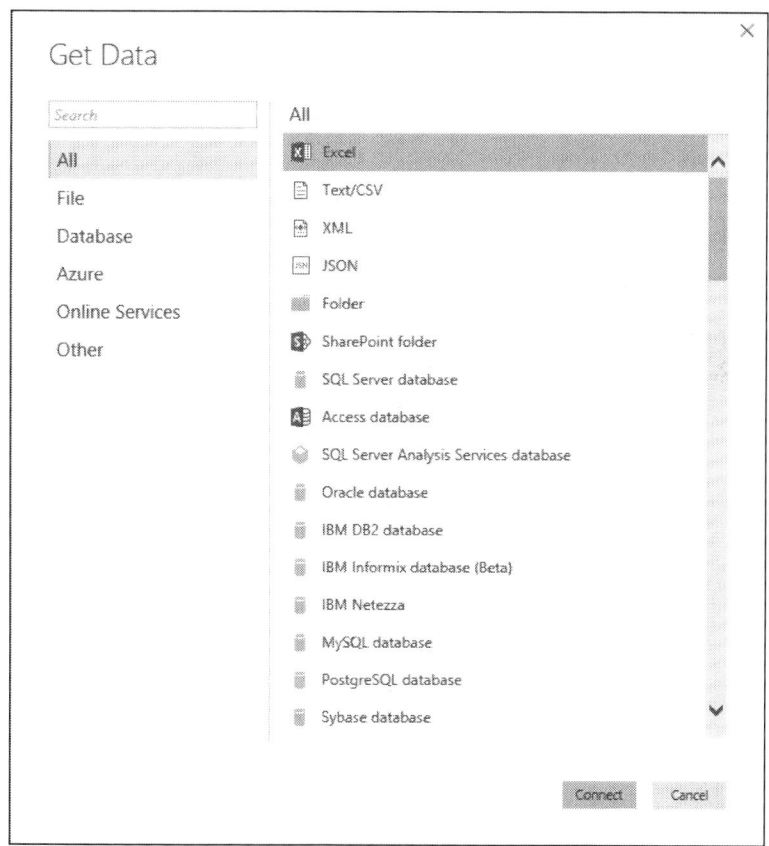

Figure 1.12: The Get Data dialog box

The **Open** dialog box appears.

11. Navigate to the location where the file to be connected to Power BI is located.
12. Select the file that you want to connect to Power BI.
13. Click the **Open** button, as shown in Figure 1.13.

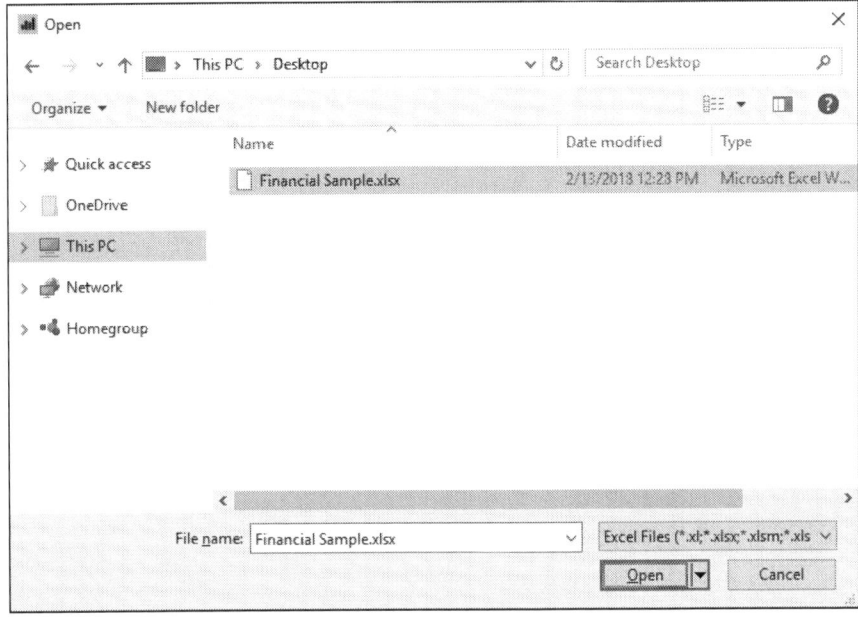

Figure 1.13: The Open dialog box

After clicking the **Open** button, Power BI connects to the selected data source. The **Navigator** dialog box appears.

14. Select the desired sheet under the **Display Options** section to load it to Power BI. After selecting a sheet, its content appears in the right pane.
15. Click the **Load** button to load the selected sheet to Power BI, as shown in Figure 1.14.

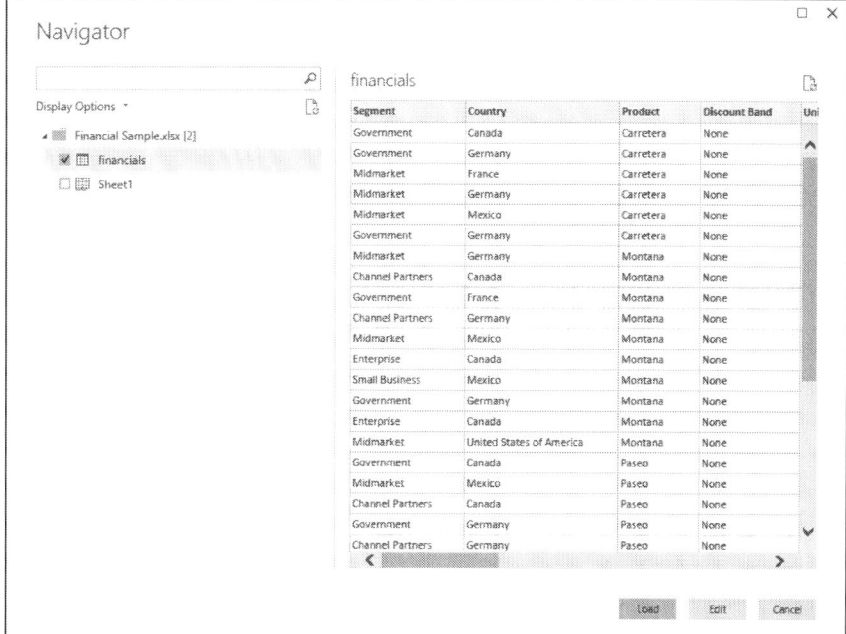

Figure 1.14: Loading a sheet

The selected sheet is loaded to Power BI and the **FIELDS** pane appears displaying the fields of the sheet, as shown in Figure 1.15.

Figure 1.15: The FIELDS pane

By default, you get the Report view of Power BI Desktop. You can access the Data view by clicking the Data icon in the left pane. It displays all the data in tabular form, as shown in Figure 1.16.

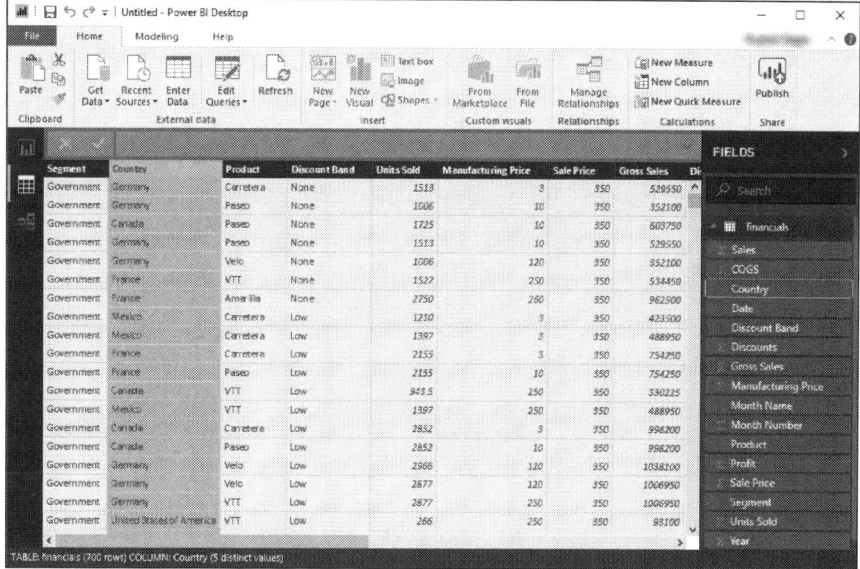

Figure 1.16: Displaying the Data view

Shaping Data or Data Modeling

Power BI Desktop allows users to connect to several data sources. It also allows users to shape the data according to their requirements. Shaping data is the same thing as transforming data. Some of the tasks associated with shaping data include:

- Modifying columns or tables
- Modifying data types
- Adding or removing rows/columns
- Making the first row as header

To shape or transform data, Power BI Desktop provides the Query Editor window. This window allows you to shape data and create relationships. You can open the Query Editor window by clicking the **Edit Queries** button under the **External Data** section of the **Home** tab of Power BI Desktop. Figure 1.17 shows the **Query Editor** window.

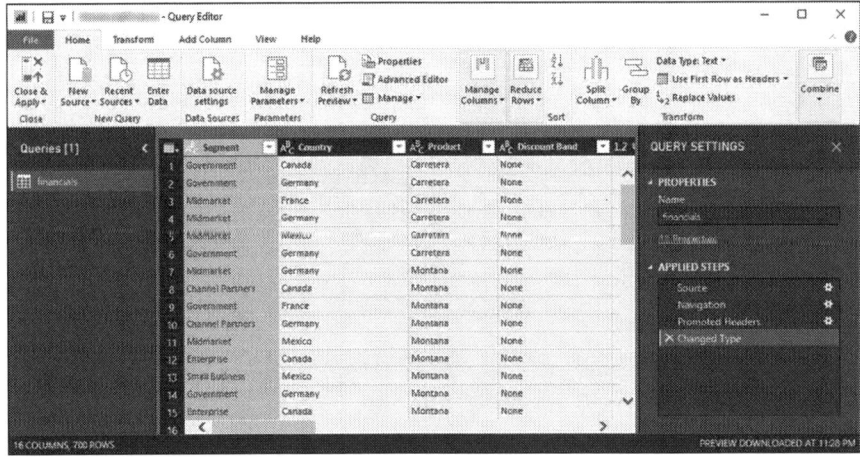

Figure 1.17: The Query Editor window

Note

The Query Editor window displays a blank pane when there is no data source.

Let us go through the user interface of the Query Editor window before providing a detailed description of shaping data. Usually, the Query Editor window provides tools tabs within the Ribbon. The Ribbon provides a large set of commands or options segregated into different tabs as outlined below.

1. **Home tab:** This tab contains the query related commands, such as adding a new data source, entering data, managing columns, etc. Figure 1.18 shows the Home tab.

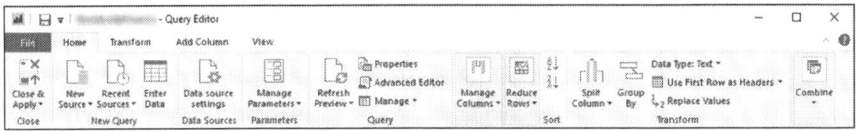

Figure 1.18: Home tab

2. **Transform tab:** This tab contains commands related to data transformation tasks, as shown in Figure 1.19.

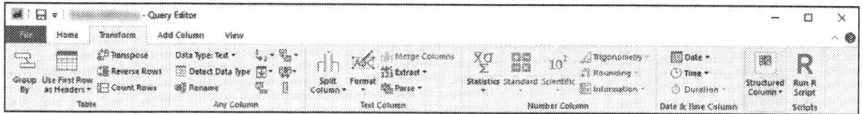

Figure 1.19: Transform tab

3. **Add Column tab:** This tab contains commands related to adding a new column, adding a custom column, and formatting a column, etc., as shown in Figure 1.20.

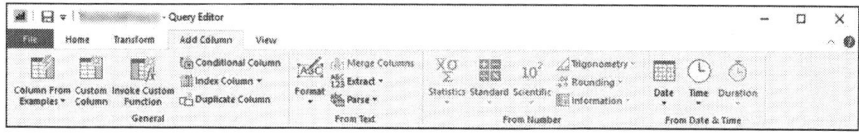

Figure 1.20: Add Column tab

4. **View tab:** This tab contains commands to open or display panes or windows, as shown in Figure 1.21.

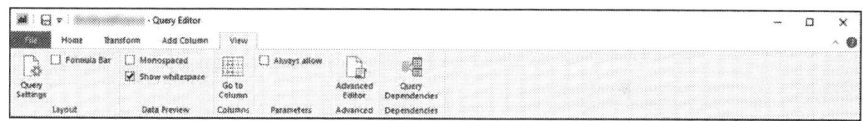

Figure 1.21: View tab

In addition to the four main tabs on the Ribbon, the window contains the following panes or sections:

1. **Queries pane:** This is the left pane of the Query Editor window, which displays the number of active queries with their names, as shown in Figure 1.22.

Figure 1.22: Queries pane

2. **Data pane:** This is the center pane where the actual data of the query displays, as shown in Figure 1.23.

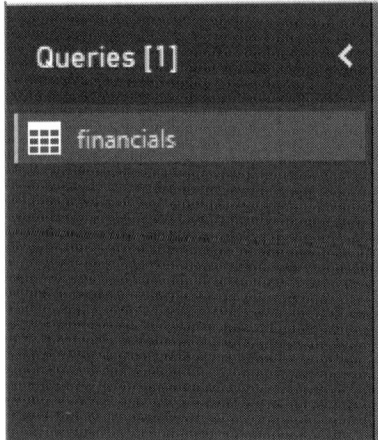

Figure 1.23: Data pane

3. **QUERY SETTINGS pane:** This pane lists the steps performed on a query under the APPLIED STEPS section, as shown in Figure 1.24.

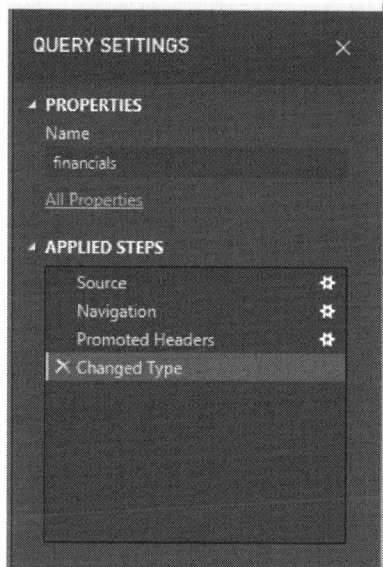

Figure 1.24: QUERY SETTINGS pane

Perform the following steps to shape data:

1. Click the **Edit Queries** button under the **External data** section of the **Home** tab of Power BI Desktop. The **Query Editor** window appears.

2. Right-click the query in the **Queries** pane (left side of window). A context menu appears.

3. Select the **Reference** option from the context menu to create the query's reference, as shown in Figure 1.25.

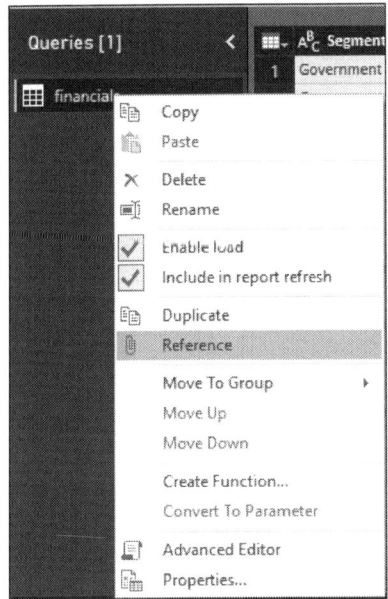

Figure 1.25: Creating a reference of a query

A reference to the selected query is created.

4. Enter the desired name for the query in the **Name** text box under the **PROP-ERTIES** section of the **QUERY SETTINGS** pane (right side of window). The specified name is updated in the **Queries** pane automatically.

5. Select the **Transform** tab. Commands related to the selected tab appear.

6. Click the **Group By** button under the **Table** section, as shown in Figure 1.26.

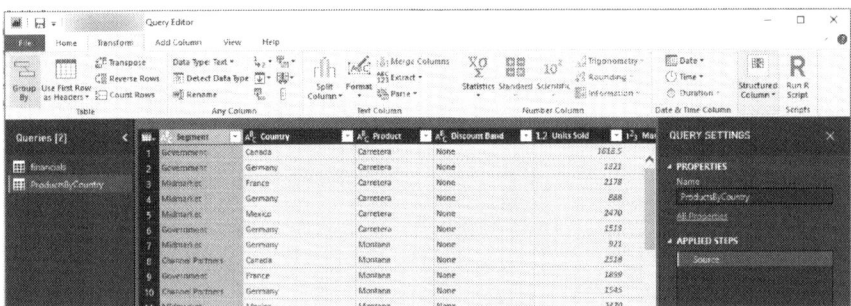

Figure 1.26: Clicking the Group By button

The **Group By** dialog box appears.

7. Select the **Basic** radio button to create a simple query, or the **Advanced** radio button to create a more complicated query.

8. Select the desired option from the **Group by** drop-down list based on how you want to apply grouping. In our case, we have selected **Country**.

9. Click the **Add grouping** button to apply another grouping.

10. Select the desired option from the **Group by** drop-down list based on how you want to apply grouping. In our case, we have selected **Product**.

11. Type the desired name of the column in the **New column name** text box. In our case, we have typed **"Total Products Sold"**.

12. Select the desired option from the **Operation** drop-down list. In our case, we have selected **Sum**.

13. Select the desired option from the **Column** drop-down list. In our case, we have selected **Units Sold**.

14. Click the **OK** button, as shown in Figure 1.27.

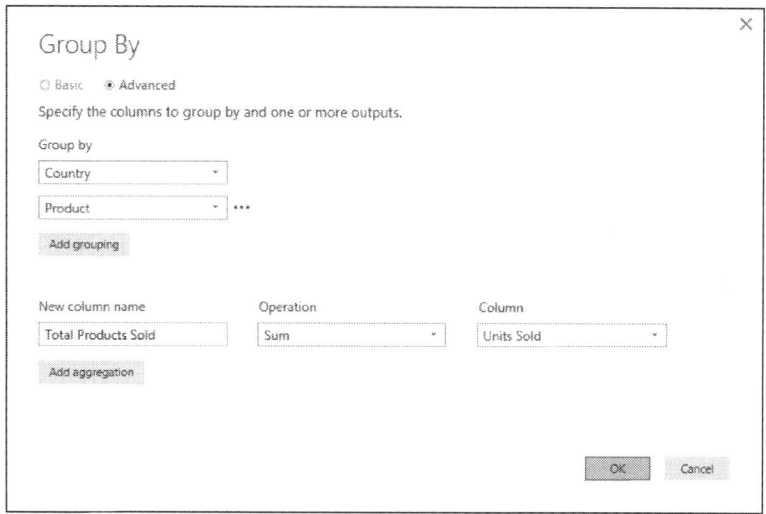

Figure 1.27: The Group By dialog box

The result of the specified query displays, as shown in Figure 1.28.

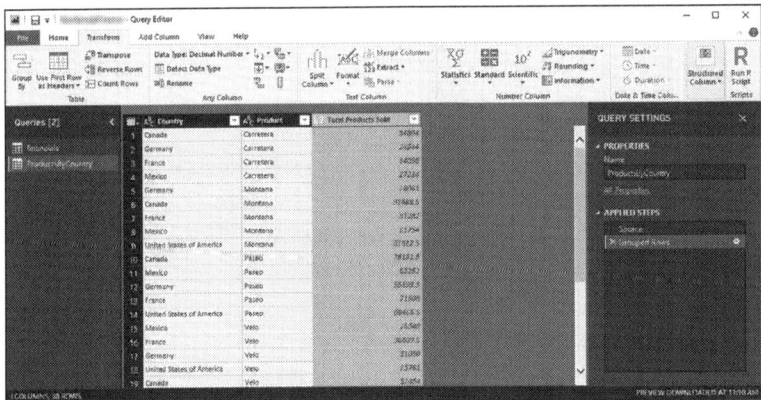

Figure 1.28: Displaying the result

15. Click the **Close & Apply** button under the **Close** section of the **Home** tab to apply the changes, as shown in Figure 1.29.

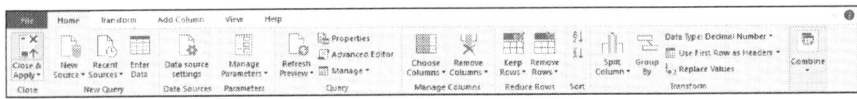

Figure 1.29: Applying changes

The changes are applied and a new table is generated. You can do additional tasks with the Query Editor window as per your requirements.

Creating Visuals

A report is all about visuals. Power BI supports a large collection of visuals, which can be used for representing the required information in more appealing manner.

Perform the following steps to create visuals based on the query created in the previous section:

1. Click the **Slicer** icon in the **VISUALIZATIONS** pane to add a slicer visual. The selected visual appears in the work area.

2. Drag the desired field from the **FIELDS** pane to the **Field** section of the **VISU-ALIZATIONS** pane. In our case, we have selected **Country**. A list of available countries appears in the slicer, as shown in Figure 1.30.

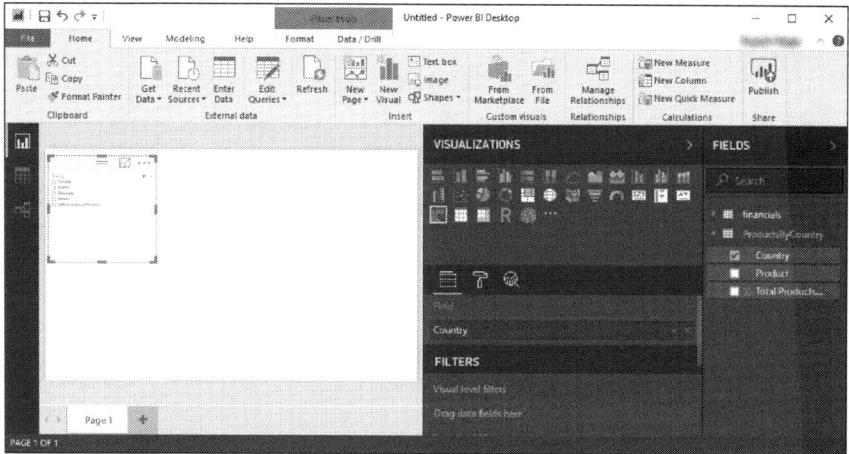

Figure 1.30: Dragging fields

3. Click the **Format** icon to view the formatting settings. The formatting settings related to the selected visual appear under each section.

4. Click the desired section to view the related settings and modify them as per your requirements.

5. Add another slicer from the **VISUALIZATIONS** pane to the work area.

6. Drag the **Product** field from the **FIELDS** pane to the **Field** section of the **VISUALIZATIONS** pane. A new slicer is added to the work area.

7. Apply the formatting settings as per your requirements.

8. Add a pie chart from the **VISUALIZATIONS** pane to the work area.

9. Drag the **Country** field in the **FIELDS** pane to the **Legend** section of the **VISUALIZATIONS** pane.

10. Drag the **Product** field in the **FIELDS** pane to the **Details** section of the **VISU-ALIZATIONS** pane.

11. Drag the **Total Products Sold** field in the **FIELDS** pane to the **Values** section of the **VISUALIZATIONS** pane, as shown in Figure 1.31.

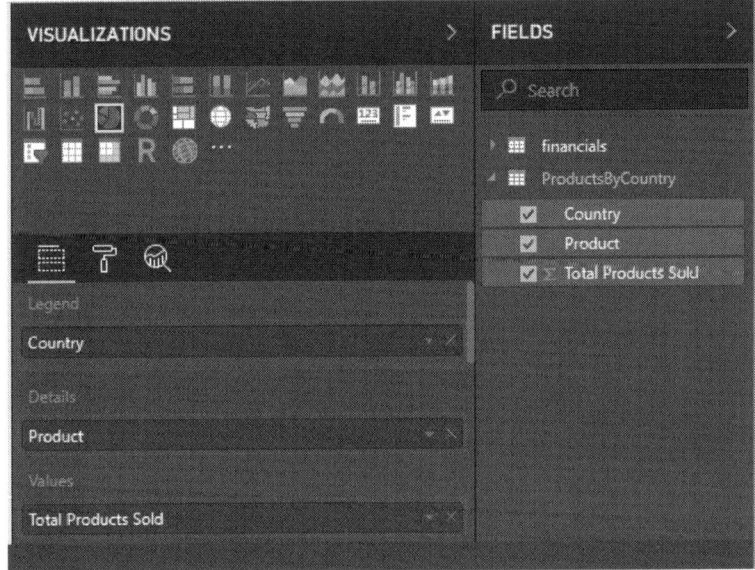

Figure 1.31: Dragging fields

12. Apply the formatting settings as per your requirement.

The visuals appear in the work area, as shown in Figure 1.32.

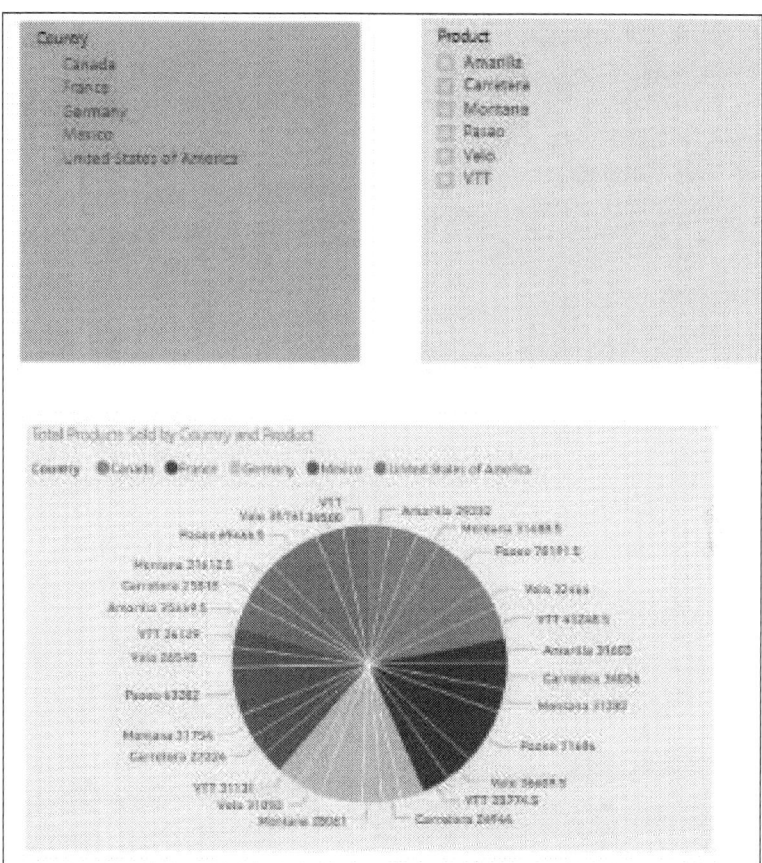

Figure 1.32: Displaying the visuals

Saving the Report

Generally speaking, it is good practice to save a report regularly.

Note

The extension of Power BI report is .pbix.

Perform the following steps to save the report:

1. Select the **File** tab from the **Ribbon**. The Backstage View appears.
2. Select the **Save** option from the Backstage View.

The **Save As** dialog box appears.

3. Select the location wherein you want to save the report.
4. Type the desired name for the report in the **File name** field. In our case, we have typed **DemoReport**.
5. Click the **Save** button, as shown in Figure 1.33.

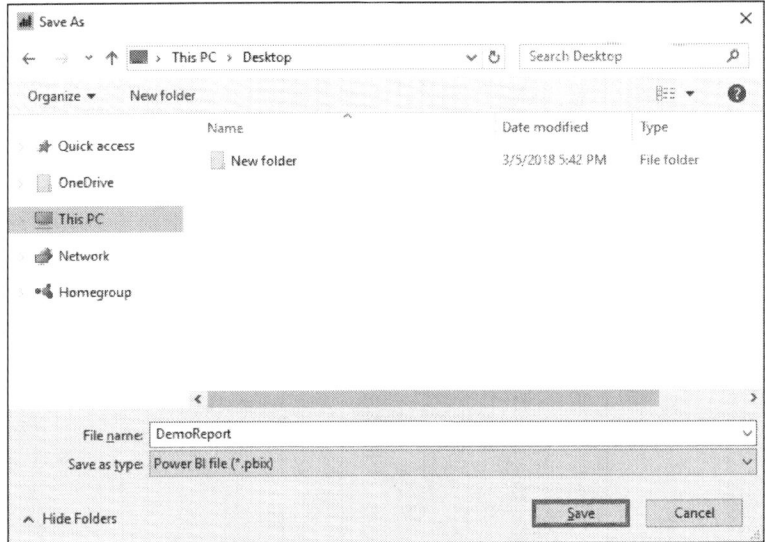

Figure 1.33: Saving a report

The report is saved with the specified name.

Power BI Service

Power BI Service/Power BI Online is a business intelligence service that hosts reports in the cloud (Microsoft Azure). The major difference between Power BI Desktop and Power BI Service is that the former focuses on creating data while the latter focuses on sharing data. Another difference between the two lies in their interface.

Note

You should have the sign-in credentials already to use Power BI Service. However, you can sign up for free if you do not have the credentials.

User Interface of Power BI Service

Figure 1.34 shows the user interface of Power BI Service.

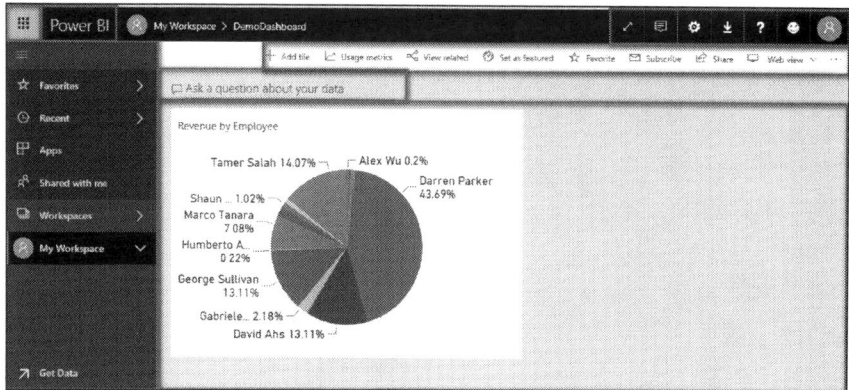

Figure 1.34: User interface of Power BI Service

A description of user interface components shown in Figure 1.34 is as follows:

1. **Navigation pane:** Allows users to navigate between workspaces and the Power BI building blocks such as dashboards, reports, workbooks, and datasets. Figure 1.35 shows the navigation pane.

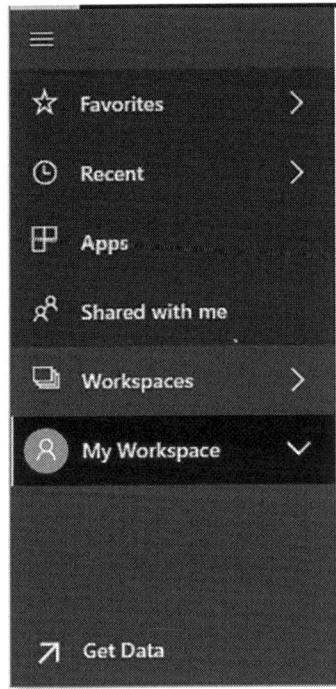

Figure 1.35: Navigation pane

The navigation pane contains several options. A brief description of these options is as follows:

- **Expand/collapse icon (▆):** Is used to expand/collapse the navigation menu.
- **Favorites:** Is used to open or manage the favorite content.
- **Recent:** Is used to view and open the content visited recently.
- **Apps:** Is used to view, open, or delete an app.
- **Shared with me:** Is used to view and search the content shared by a colleague/friend with you.
- **Workspaces:** Displays available workspaces.
- **Get Data button:** Is used to add datasets, reports, and dashboards to Power BI.

2. **Canvas:** Is a collection of tiles and as such, provides a visualization area for tiles/reports. It displays a report page when we open the report editor. A tile can be added to a dashboard. The act of adding tiles to the dashboard is known as pinning. Figure 1.36 shows the canvas.

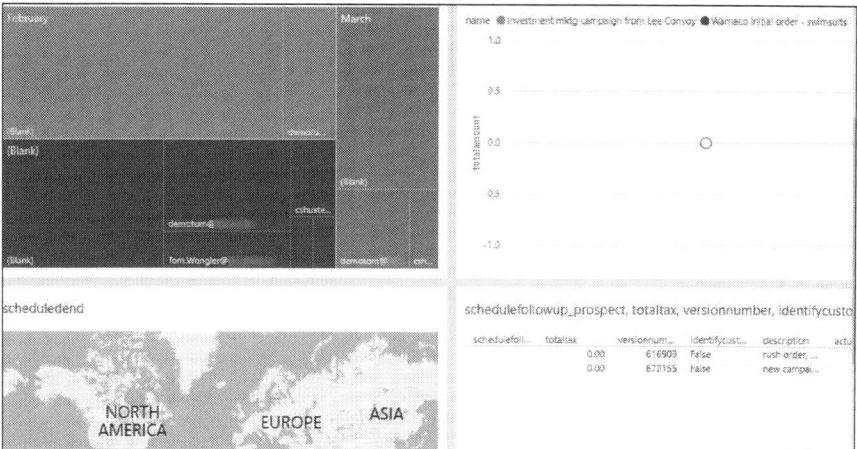

Figure 1.36: Canvas

3. **Q&A box:** Looks for the solution in the dataset linked to the dashboard and provides an answer to your query in the form of visualization. It can also be used to add content to the dashboard. Figure 1.37 shows the Q&A box.

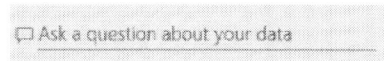

Figure 1.37: Q&A box

4. **Icon buttons:** Provide help in performing specific tasks, such as opening dashboard in full-screen mode, viewing notifications, applying different settings, viewing downloads, getting help, and sending feedback. Figure 1.38 shows the icon buttons.

Figure 1.38: Icon buttons

5. **Dashboard title:** Displays the name of the workspace preceded by the dashboard/report name. Each section in the dashboard title acts as an active link. Figure 1.39 shows the dashboard title.

My Workspace 〉 DynamicCRMDashboard

Figure 1.39: Dashboard title

6. **Office 365 app launcher:** Helps in easily locating and opening all Office 365 apps. When you click the Office 365 app launcher icon (▦), a list of Office 365 apps appears, as shown in Figure 1.40.

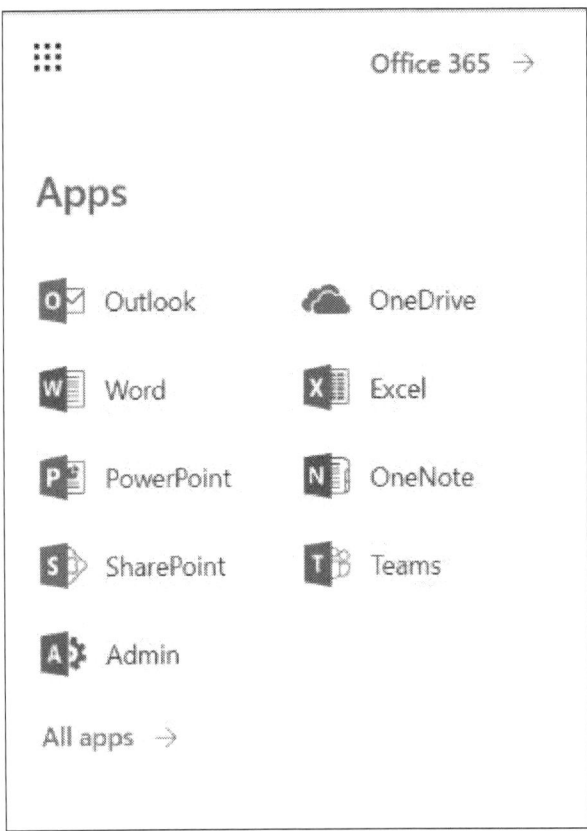

Figure 1.40: List of Office 365 apps

These apps help users quickly launch their emails and documents, etc.

7. **Power BI Home button:** Opens the featured dashboard if you have set it. Otherwise, it opens the viewed dashboard. Figure 1.41 shows the Power BI Home button.

Figure 1.41: Power BI Home button

8. **Labeled icon buttons:** Allow users to interact with the content. You can select the Ellipsis icon (...) to see more options for different tasks including duplicating, printing, and refreshing the dashboard, etc. Figure 1.42 shows the labeled icon buttons.

Figure 1.42: Labeled icon buttons

Building Blocks of Power BI Service
The following are some major building blocks of Power BI:
- Dashboards
- Reports
- Workbooks
- Datasets

These building blocks are grouped into workspaces, wherein workspaces act as containers for these building blocks. In Power BI, the following two types of workspaces are available:
- **My workspace:** Allows a Power BI user to work with his/her content personally. No one can access your "My workspace". To share the content with others, you need to create an App workspace (outlined below) and bundle the required content in the app, and share it with others in your organization.
- **App workspaces:** Allow users to share content with others as well as create, publish, and control apps. App workspaces are considered the major content containers, which help in building a Power BI app.

Dashboards
A dashboard is a collection of tiles, from no tiles to any number of tiles. A tile is the representation of data that is pinned to the dashboard. A tile can be created through several components including reports, datasets, dashboards, Excel,

SSRS, etc. You can use the Add tile button to create individual tiles. You can add text boxes, videos, streaming data, etc., to a tile directly from the dashboard.

In addition to pinning a tile from a report, you can pin the entire report pages as a single tile to the dashboard.

You can select the Dashboards tab to view the available dashboards associated with the workspace you selected. You can simply select a specific dashboard to open it and view its content. Additionally, each dashboard provides a personalized view of the associated datasets. You should note that you can not make editing changes to the dashboards and reports shared by others. However, you can edit the datasets as well as reports if you own the dashboard. Figure 1.43 shows a list of dashboards available under the Dashboards tab.

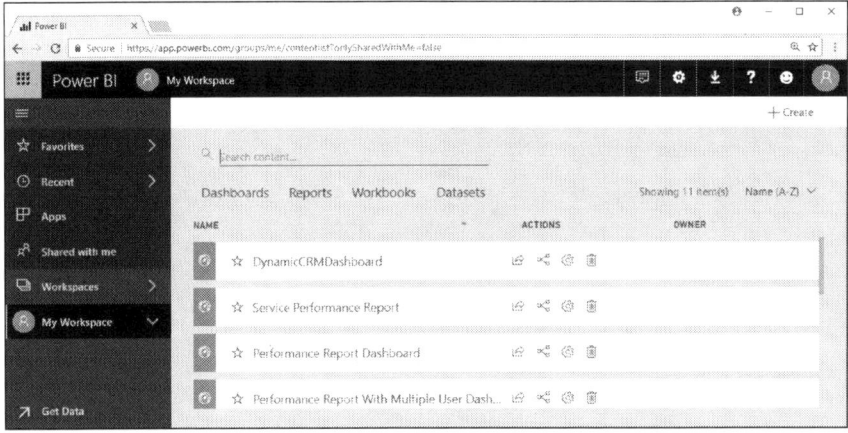

Figure 1.43: Dashboards

You may need dashboards for several reasons. Some of them are listed as follows:
- View decision-making information in one glance.
- Control important information about the organization.
- Ensure that the same page and information is accessible and uniform to all colleagues.
- Administer the business condition or product condition.
- Build a customized view of a dashboard.

Reports
A Power BI report is a collection of visualizations/visuals such as charts and graphs. These visuals are taken from a single dataset. There are several ways to create reports. Some of them are as follows:

- Create reports from scratch within Power BI.
- Import reports with dashboards shared by colleagues.
- Create reports by connecting datasets to tools including Excel, Power BI Desktop, databases, SaaS applications, and Apps. For example, you can create a report based on the Power View sheets available in an Excel workbook when a user is connected to it.

To interact with reports, Power BI provides the following two views:
1. **Reading view:** Is the default view of a report. This view allows both owners as well as recipients of the shared link to access the report and its content. However, editing can not be performed in this view.
2. **Editing view:** Allows only owners, co-owners, and granted users to explore, design, and modify reports. You can open a report in the Editing view by clicking the Edit report button in the labeled icon buttons.

All reports associated with a workspace appear under the Reports tab, as shown in Figure 1.44.

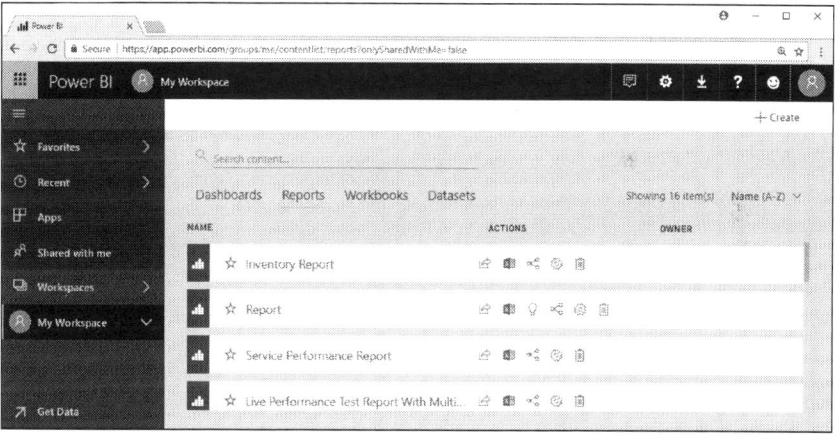

Figure 1.44: Displaying reports

Datasets
A dataset is defined as a set of data that a user imports or connects to in Power BI. Power BI can connect to all types of datasets and collates them together.

The following are important points related to datasets:
- Datasets are linked with workspaces wherein a single dataset can be linked to multiple workspaces.
- A dataset can be linked to different reports.
- Visualizations related to a dataset can be viewed on different dashboards.

All datasets associated with a workspace appear under the Datasets tab, as shown in Figure 1.45.

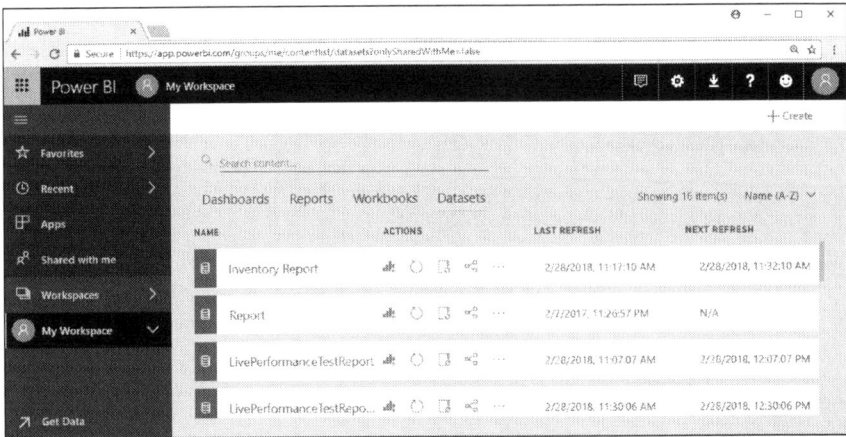

Figure 1.45: Displaying datasets

Workbooks
A workbook is a special type of dataset that can be imported or connected to Power BI. When you select Excel as an option for "Get data" and click the Connect button, the associated workbook appears in Power BI. From here, you can pin elements directly to the dashboard.

Publishing a Report

Publishing a report created in Power BI Desktop to Power BI Service is simple. This can help users to easily access the report.

Perform the following steps to publish a report:
1. Open the report in Power BI Desktop.

2. Click the **Publish** button under the **Share** section of the **Home** tab, as shown in Figure 1.46.

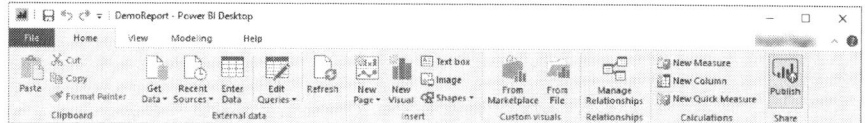

Figure 1.46: Publishing a report

The **Publish to Power BI** dialog box appears.
3. Select the desired destination from the **Select a destination** list box.
4. Click the **Select** button, as shown in Figure 1.47.

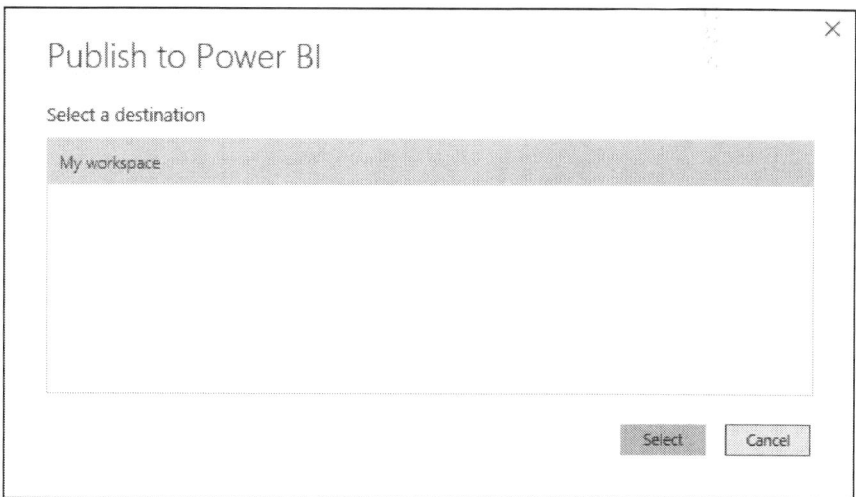

Figure 1.47: The Publish to Power BI dialog box

The **Publishing to Power BI** dialog box displays the status of publishing the report to Power BI. Once the publishing is successful, you can open the report in Power BI by clicking the **Open 'DemoReport.pbix' in Power BI** link, as shown in Figure 1.48.

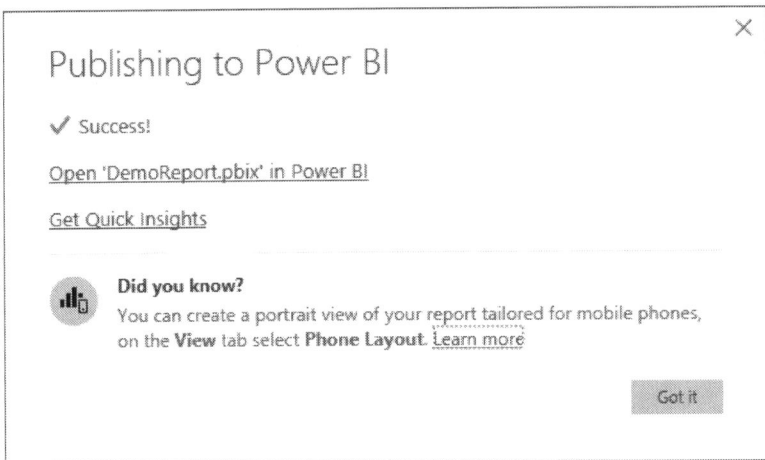

Figure 1.48: The Publishing to Power BI dialog box

The published report appears in Power BI Service, as shown in Figure 1.49.

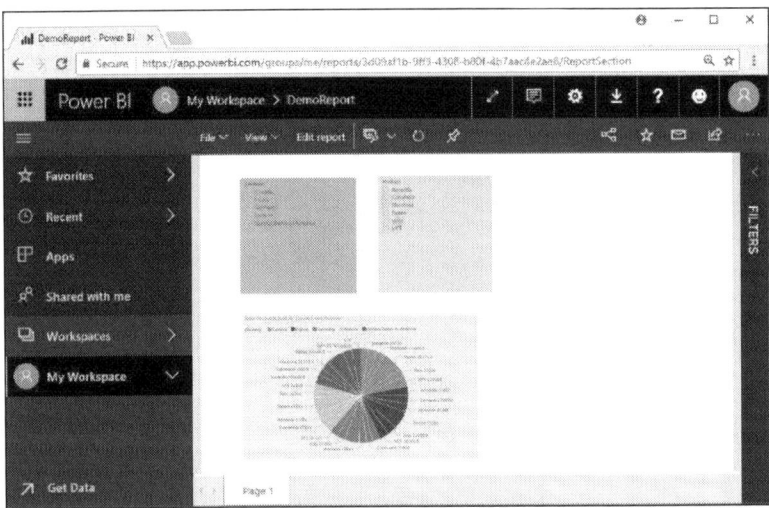

Figure 1.49: Displaying the published report

Summary

This chapter provided a brief overview of one of the prominent data visualization tools, that is, Power BI. Power BI is a data visualization and business intelligence tool that allows business owners to easily obtain their business statistics. The advanced features associated with Power BI make it an ideal choice for business analytics executives. This chapter also provided detailed information about Power BI Desktop and Power BI Service. You also learned to create and publish a report to Power BI Service.

Chapter 2
Power BI Azure Application

Power BI can be integrated with Azure services to obtain real-time insights into your business. With the help of Azure APIs, you can see real-time business data in an intuitive manner irrespective of the nature of processing of the data.

You can create reports in Power BI and embed them into a Web app. You can also view real-time streaming of data, which means a live update to data can be seen in visuals pinned to Power BI dashboards.

Embedding Power BI Reports into a Web App

Power BI reports and dashboards can be embedded into a Web app with the help of APIs and sample code available from a source repository like Git.

Perform the following steps to embed Power BI reports into a Web app:
1. Build the source code for Web app in Visual Studio.
2. Register your app for Power BI by performing the following steps:
   a. Navigate to the following link:
      https://dev.powerbi.com/apps

   The **Register an Application for Power BI** window appears.
   b. Login to your Power BI account. Once the login is successful, you are redirected to the **Register an Application for Power BI** window.
   c. Enter the desired name for your app in the **App Name** text box.
   d. Select the desired app type from the **App Type** drop-down list, as shown in Figure 2.1.

DOI 10.1515/9781547400720-002

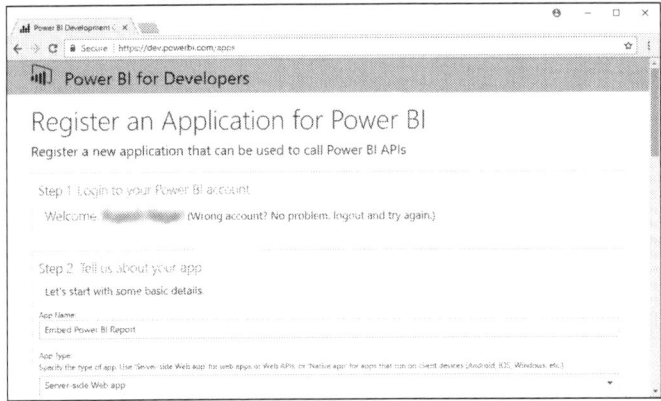

Figure 2.1: Specifying app details

e. Specify the redirect URL in the **Redirect URL** text box.

f. Specify the home page URL in the **Home Page URL** text box.

g. Select the desired APIs to access under the **Dataset APIs**, **Report and Dashboard APIs**, and **Other APIs** sections, as shown in Figure 2.2.

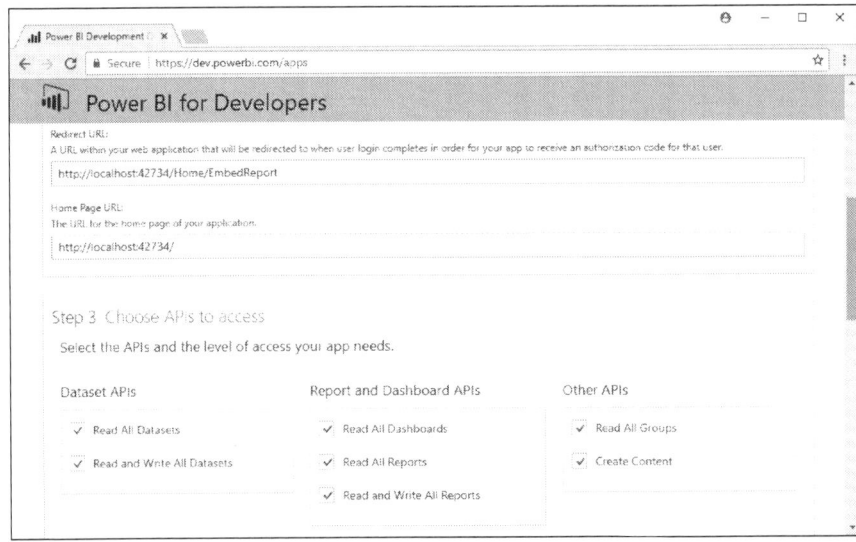

Figure 2.2: Selecting APIs to access

h. Click the **Register App** button under the **Register your app** section to register the app, as shown in Figure 2.3.

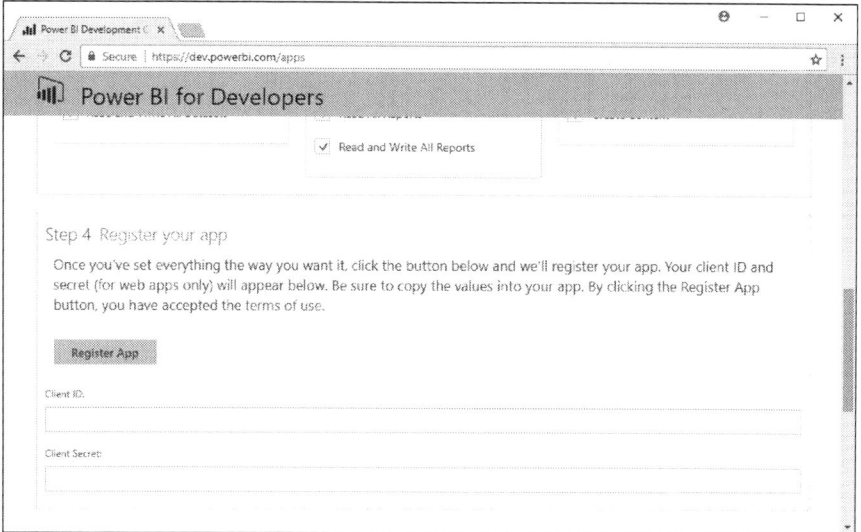

Figure 2.3: Registering the app

As you click the **Register App** button, client ID and client secret values are generated in the **Client ID** and **Client Secret** fields, respectively.

3. Open the **Cloud.config** file and enter values from the **Client ID** and **Client Secret** fields in the **ClientID** and the **ClientSecret** settings, as shown in Figure 2.4.

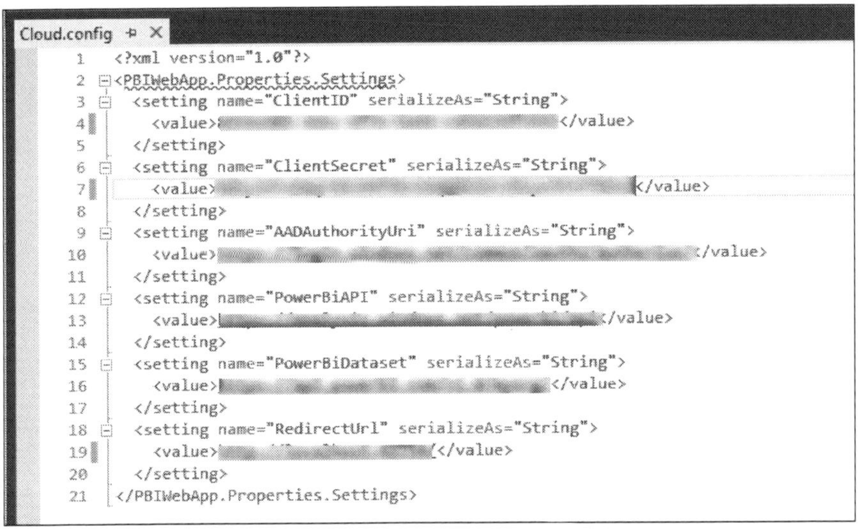

Figure 2.4: Specifying values in the ClientID and ClientSecret settings

4. Specify the report index in the GetReport() function to embed, as shown in Figure 2.5.

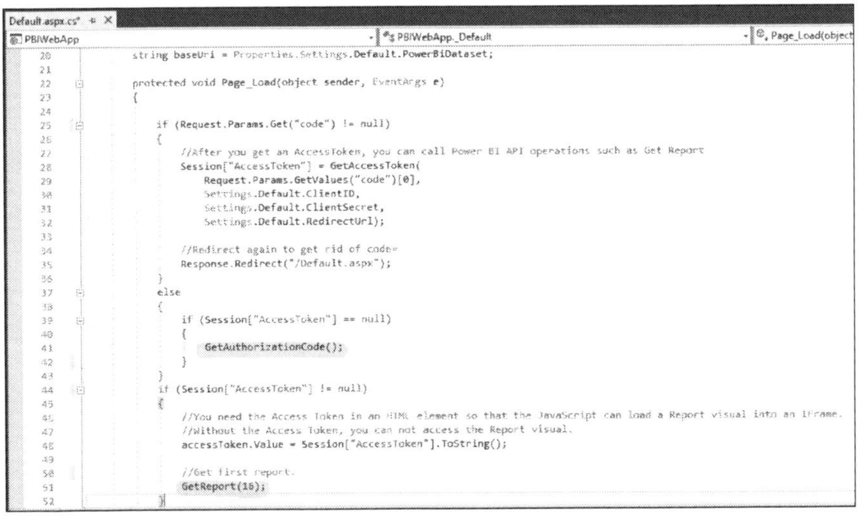

Figure 2.5: Specifying the report index

5. Press the **F5** key on the keyboard to run the application.

The specified report is embedded into the Web app, as shown in Figure 2.6.

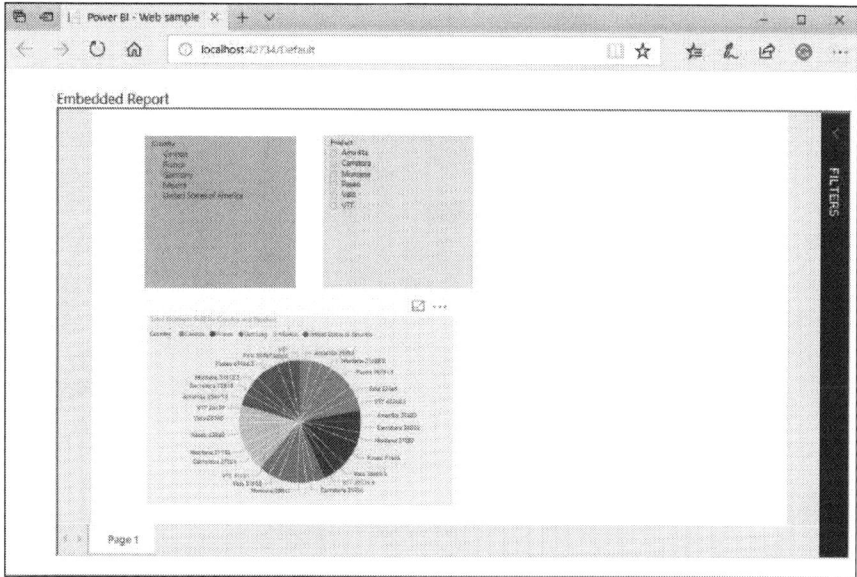

Figure 2.6: Embedding a report

Real-Time Streaming

Real-time streaming in Power BI helps users in streaming data in Power BI dashboards in real time, which means that an update to data is immediately available in visuals pinned to Power BI dashboards. You can collect real-time information from different sources that provide time-sensitive data. To visualize real-time data, you need to set up the real-time streaming dataset in Power BI.

Exploring Real-Time Streaming Datasets

Power BI supports three real-time streaming datasets, which are as follows:
1. **Push dataset:** This dataset pushes data into Power BI Service, which automatically creates a database at the time of creating the dataset. This database stores the data in a queue format to allow users to create reports based on the available data. You can pin the visuals of the created report to the dashboard, which are updated in real time when there is an update to the data.

2. **Streaming dataset:** Similar to the Push dataset, this dataset also pushes data into Power BI. However, data is stored temporarily in the cache memory, which is especially used for displaying visuals such as line charts. Some considerations related to streaming dataset are as follows:
   - The data that flows from the stream cannot be used for creating report visuals because there is no dedicated database.
   - The data can be visualized by adding a tile and using the "custom streaming data" data source.
   - It shows real-time data quickly in the custom streaming tiles.
3. **PubNub streaming dataset:** This dataset provides the Power BI web client that uses the PubNub SDK. This SDK reads the available PubNub data stream and restricts Power BI Service to store any data. Some considerations related to PubNub streaming dataset are as follows:
   - The data that flows from the stream cannot be used for creating report visuals because there is no dedicated database.
   - Several report functionalities are unavailable including report filtering and use of custom visuals, etc.
   - A PubNub data stream can be configured as the source and a tile can be added to the dashboard for visualizing the PubNub streaming dataset.
   - The tiles on the dashboard display real-time data quickly.

Different Ways of Pushing Data

You can push data into a dataset using:
- Power BI REST APIs
- Streaming Dataset UI
- Azure Stream Analytics

Viewing Real-Time Streaming of Data

Perform the following steps to view real-time streaming of data:
1. Open the Power BI Service portal.
2. Click the **My Workspace** button in the left pane.
3. Click the **Create** button in the right pane. A drop-down list appears.
4. Click the **Streaming dataset** option, as shown in Figure 2.7.

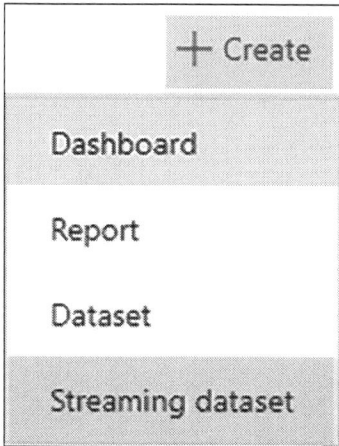

Figure 2.7: Creating a streaming dataset

The **New streaming dataset** wizard appears.

5. Select the desired data source under the **Choose the source of your data** section. In our case, we have selected **PUBNUB**.
6. Click the **Next** button, as shown in Figure 2.8.

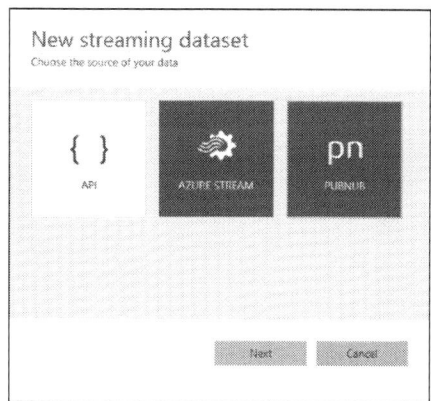

Figure 2.8: The New streaming dataset wizard

The next page appears requiring details such as the subscription key and channel name.

7. Enter the desired dataset name in the **Dataset name** text box.
8. Enter the relevant subscription key in the **Sub-key** text box.
9. Enter the relevant channel name in the **Channel name** text box.

10. Click the **Next** button, as shown in Figure 2.9.

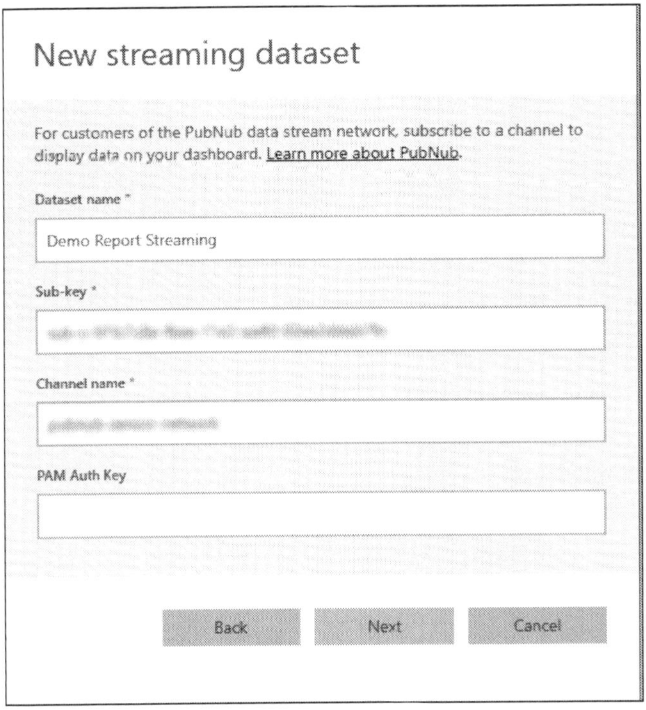

New streaming dataset

For customers of the PubNub data stream network, subscribe to a channel to display data on your dashboard. Learn more about PubNub.

Dataset name *

Demo Report Streaming

Sub-key *

Channel name *

PAM Auth Key

Back   Next   Cancel

Figure 2.9: Specifying subscription details

The next page appears requiring the values from the stream.

11. Specify values and select the relevant data type for the values under the **Values from stream** section.

12. Click the **Create** button, as shown in Figure 2.10.

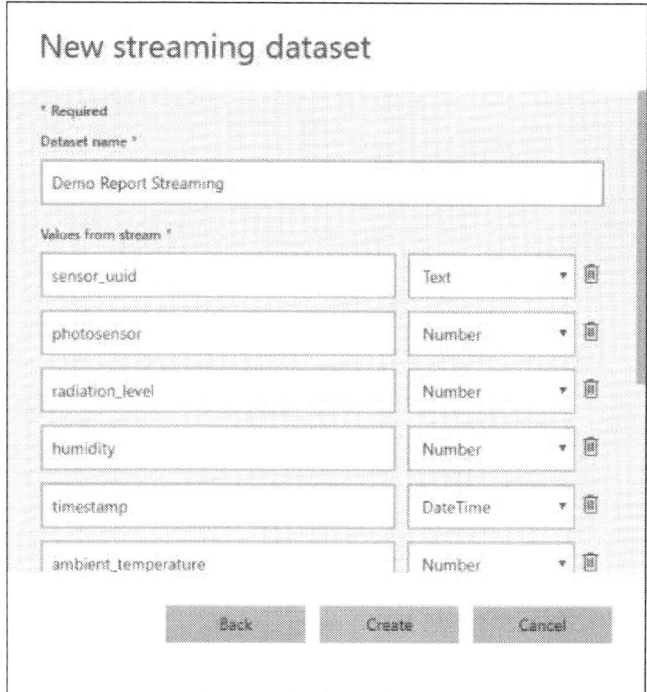

Figure 2.10: Specifying values from the stream

The specified dataset is created. You can locate this dataset under the **Datasets** tab, as shown in Figure 2.11.

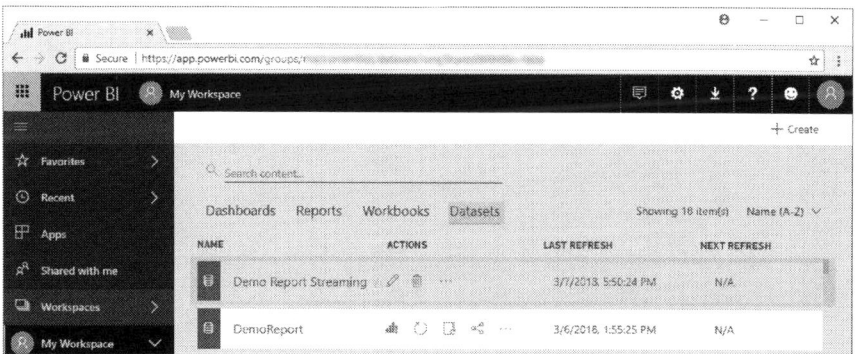

Figure 2.11: Viewing the dataset

13. Click the **Create** button. A drop-down list appears.
14. Click the **Dashboard** option to create a new dashboard, as shown in Figure 2.12.

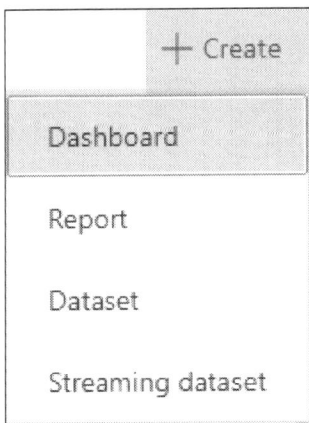

Figure 2.12: Creating a dashboard

The **Create dashboard** dialog box appears.

15. Type the desired name for the dashboard in the **Dashboard name** text box.
16. Click the **Create** button, as shown in Figure 2.13.

Figure 2.13: The Create dashboard dialog box

The dashboard is created with the specified name.

17. Click the **Add tile** button, as shown in Figure 2.14.

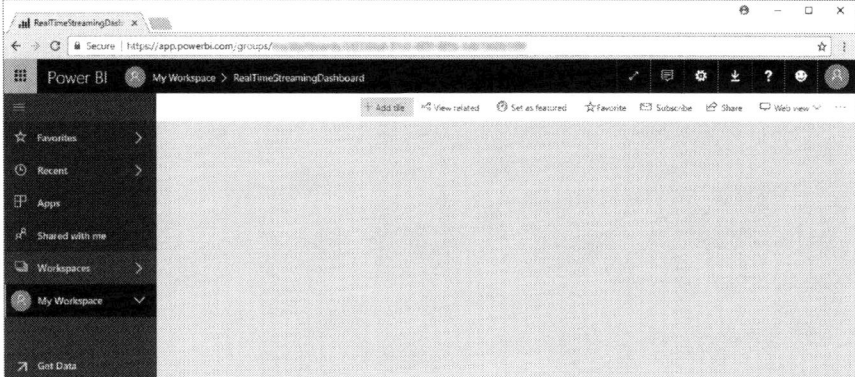

Figure 2.14: Clicking the Add tile button

The **Add tile** dialog box appears.

18. Select the **Custom Streaming Data** option under the **REAL-TIME DATA** section.

19. Click the **Next** button, as shown in Figure 2.15.

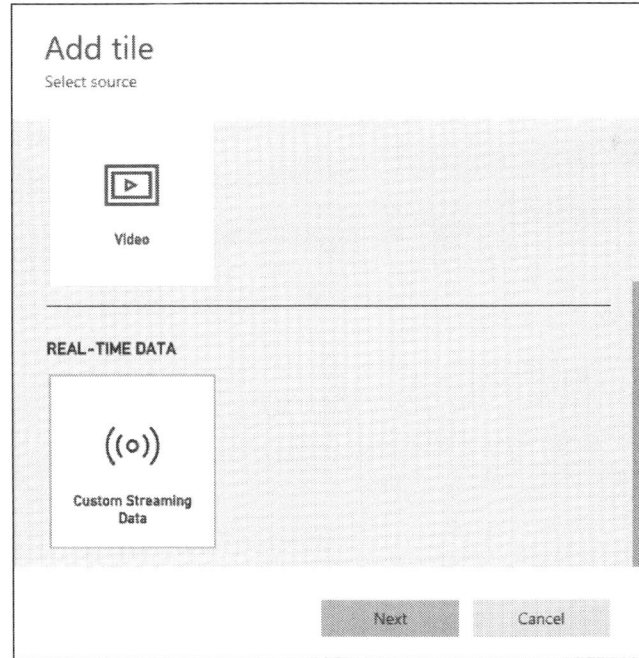

Figure 2.15: The Add tile dialog box

The **Add a custom streaming data tile** dialog box appears.

20. Select a streaming dataset. In our case, we have selected the dataset created earlier that appears under the **YOUR DATASETS** section.

21. Click the **Next** button, as shown in Figure 2.16.

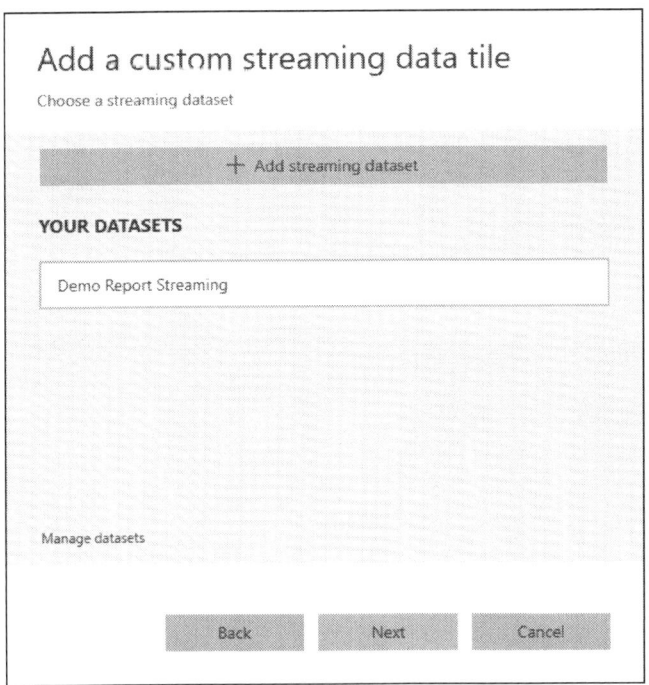

Figure 2.16: The Add a custom streaming data tile dialog box

The **Visualization design** page appears.

22. Select the desired visualization type from the **Visualization Type** drop-down list. In our case, we have selected **Line chart**, as shown in Figure 2.17.

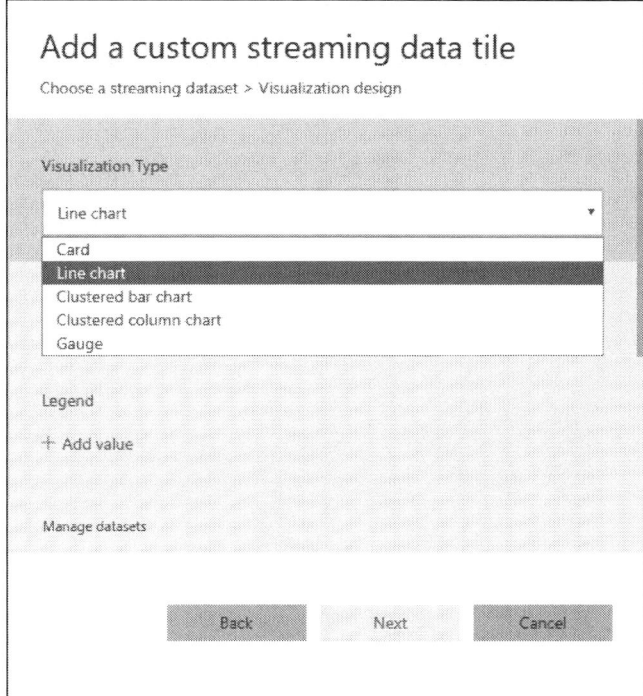

Figure 2.17: Selecting a visualization type

23. Select the desired axis value from the **Axis** drop-down list.
24. Select the desired value from the **Values** drop-down list.
25. Click the **Next** button, as shown in Figure 2.18.

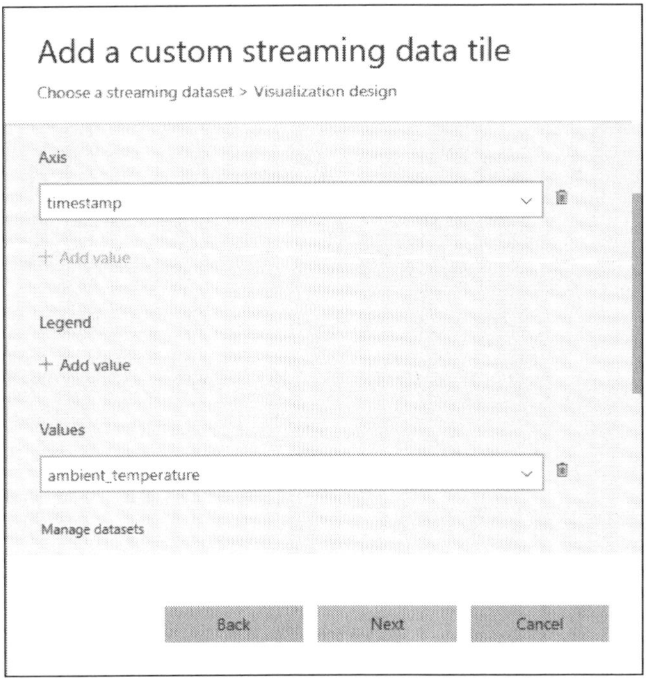

Figure 2.18: The Visualization design page

The **Tile details** dialog box appears.

26. Select the **Display title and subtitle** checkbox to display the title as well as a subtitle.
27. Enter the desired title and subtitle in the **Title** and **Subtitle** text boxes, respectively.
28. Click the **Apply** button, as shown in Figure 2.19.

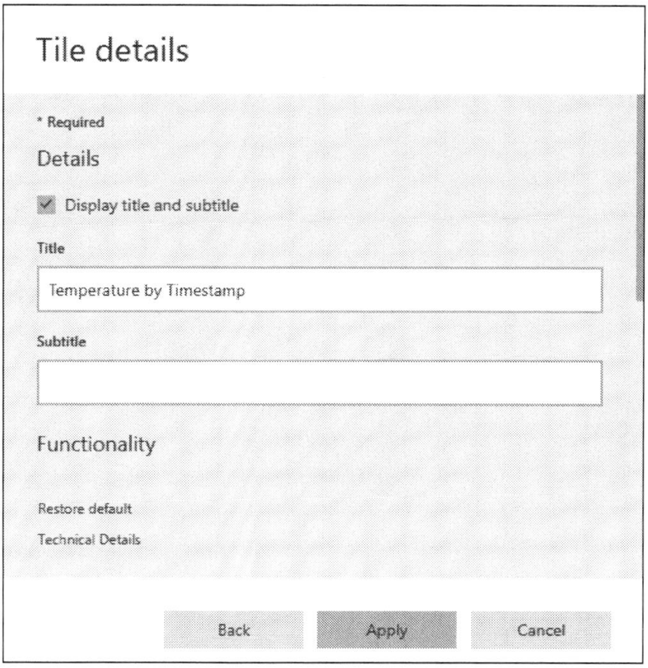

Figure 2.19: Adding tile details

The tile is added to the dashboard and real-time data appears in the tile, as shown in Figure 2.20.

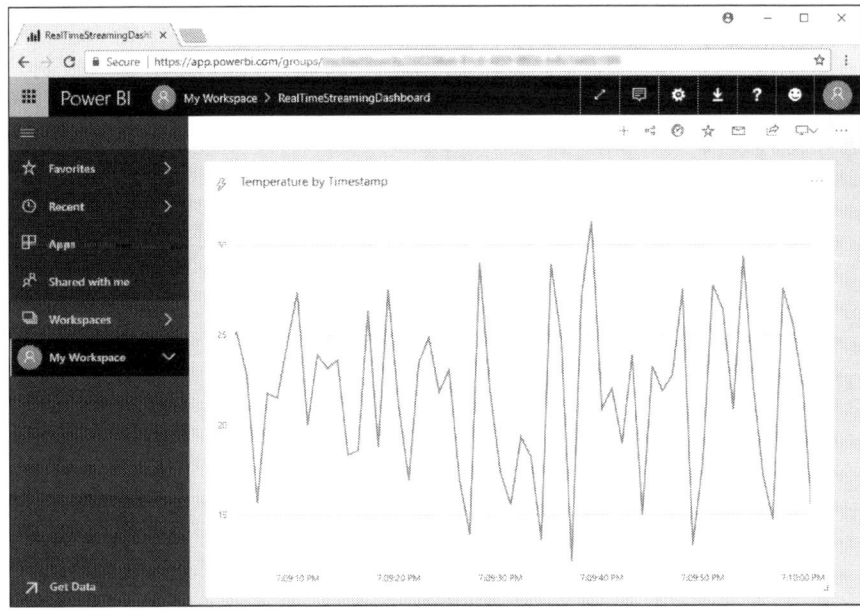

Figure 2.20: Displaying real-time streaming

Quick Insights

Power BI provides the Quick Insights feature, which applies complex algorithms to the dataset and locates different subsets of that dataset efficiently.

You can create appealing visualizations through Quick Insights by running it on either a dataset or a dashboard tile.

Perform the following steps to run Quick Insights on a dataset:
1. Open Power BI Service.
2. Select the **My Workspace** option from the left pane. A list of available dashboards appears in the right pane under the **Dashboards** tab.
3. Select the **Datasets** tab. A list of available datasets appears.
4. Click the **Ellipsis** icon (...) next to the desired dataset, as shown in Figure 2.21.

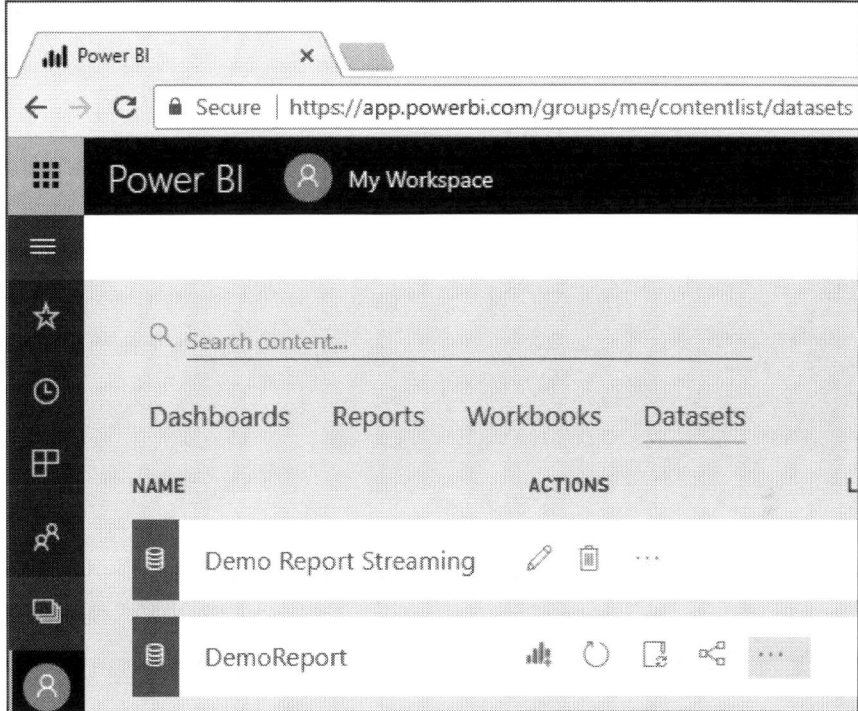

Figure 2.21: Clicking the Ellipsis icon

A drop-down list appears.

5. Click the **Get quick insights** option, as shown in Figure 2.22.

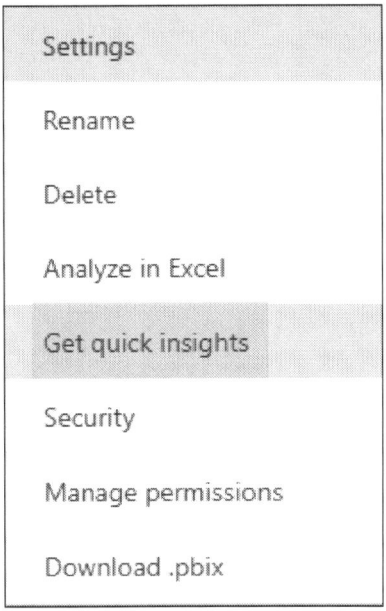

Figure 2.22: Clicking the Get quick insights option

The **Searching for insights** message box appears that displays the progress of searching insights, as shown in Figure 2.23.

Searching for insights
Searching DemoReport. We will notify you when your insights are ready.

Figure 2.23: The Searching for insights message box

The **Insights are ready** dialog box displays "**You have insights for DemoReport.**" In our example, DemoReport is the selected dataset.

6.   Click the **View insights** button, as shown in Figure 2.24.

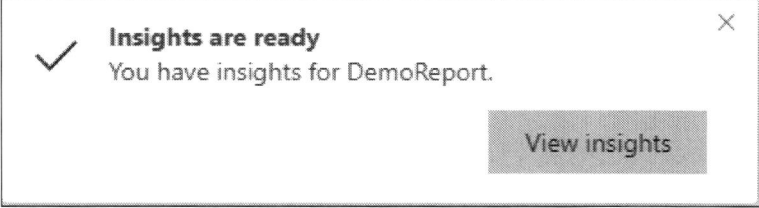

Figure 2.24: The Insights are ready dialog box

The **Quick Insights for DemoReport** window displays all possible insights for the selected dataset (DemoReport in our case), as shown in Figure 2.25.

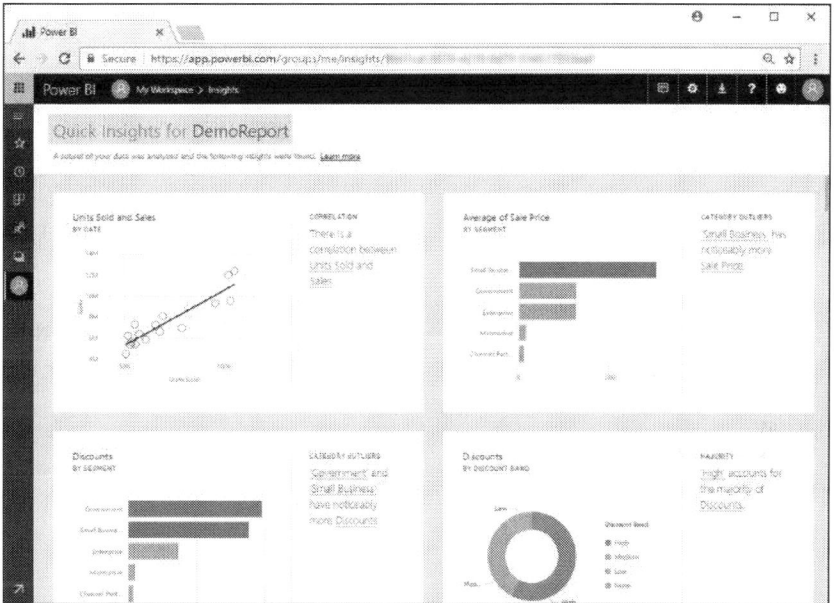

Figure 2.25: Displaying quick insights

7. Click the **Pin visual** icon ($\mathcal{P}$) to pin the selected visual to the dashboard. The **Pin to dashboard** dialog box appears.
8. Select the desired radio button to specify whether you want to pin the selected visual to an existing dashboard or a new dashboard. In our case, we have selected the **New dashboard** radio button.
9. Enter the desired name for the new dashboard.

10. Click the **Pin** button to pin the selected visual to the new dashboard, as shown in Figure 2.26.

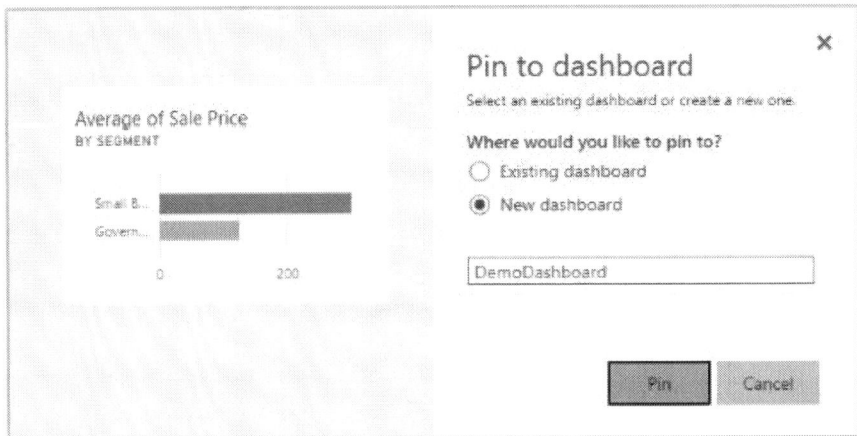

Figure 2.26: Pinning a visual to a dashboard

The **Pinned to dashboard** message box appears stating that the visualization is pinned to the specified dashboard successfully.

11. Click the **Go to dashboard** button to view the visualization pinned to the dashboard, as shown in Figure 2.27.

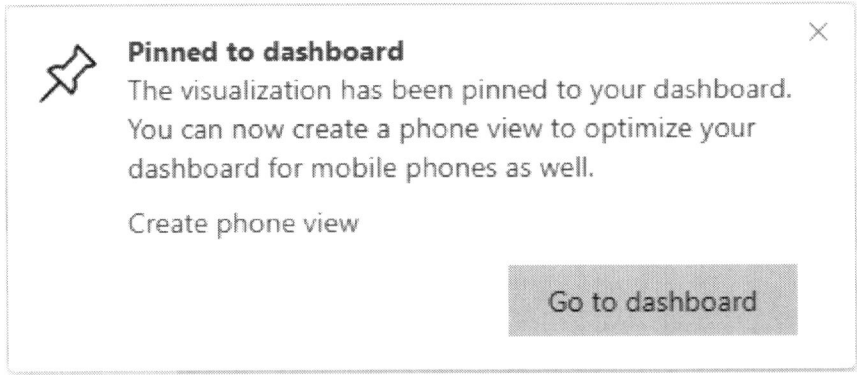

Figure 2.27: The Pinned to dashboard message box

The selected visualization is pinned to the specified dashboard.

12. Click the **Ellipsis** icon (...). A drop-down list appears.

13. Select the **Open in focus mode** option, as shown in Figure 2.28.

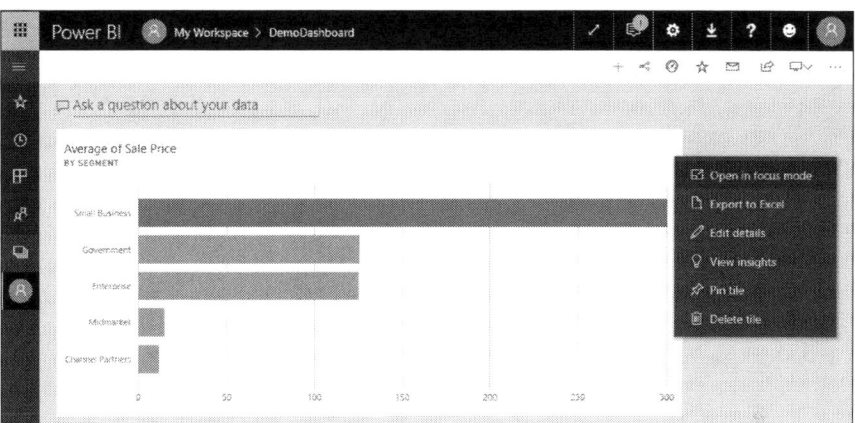

Figure 2.28: Opening the visual in focus mode

The selected visual appears in the focus mode.

14. Click the **Related Insights** icon (💡) to view the insights related to the selected visual. A list of related insights appears under the **INSIGHTS** section, as shown in Figure 2.29.

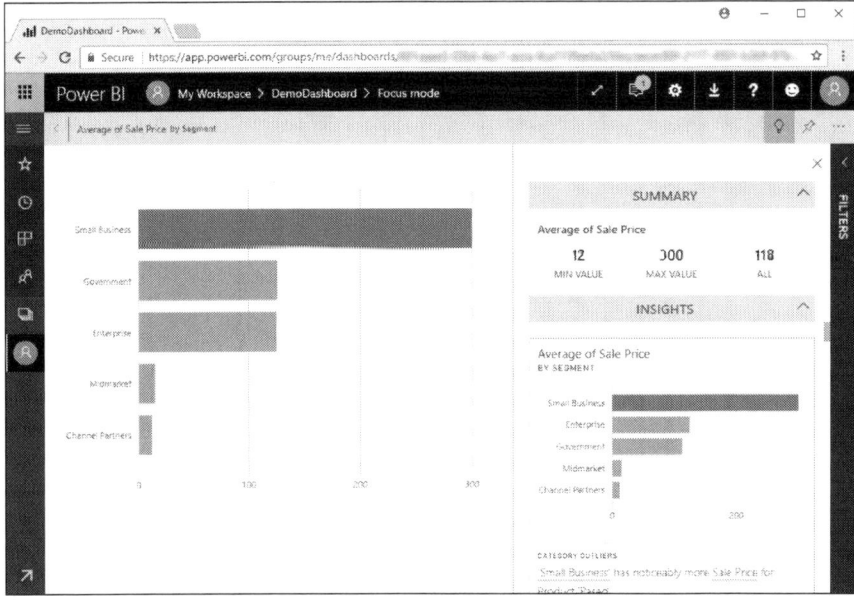

Figure 2.29: Viewing related insights

Summary

This chapter provided detailed information about embedding Power BI reports into Azure Web app to get real-time insights on a business. The steps for real-time streaming of data were covered. You learned how to register a Web app for Power BI and make necessary modifications in the code to embed Power BI reports into a Web app. Different types of real-time streaming datasets including Push dataset, Streaming dataset, and PubNub streaming dataset were discussed. You were also acquainted with the step-by-step procedure of viewing real-time streaming of data and learned how to use the Quick Insights feature of Power BI that applies complex algorithms to the dataset and quickly locates different subsets of the dataset.

Chapter 3
Power BI on Microsoft Stack

As discussed in the previous chapter, Power BI can be integrated with hundreds of data sources. One of the most prominent data sources is SQL Server, which is a relational database management system introduced by Microsoft. It is primarily used in business applications.

This chapter provides an in-depth description of integrating Power BI with SQL Server to create interactive reports.

Getting Data into Power BI from SQL Server

There are two options to bring SQL Server data into Power BI Desktop, as follows:
- Import
- DirectQuery

Using the Import Option

In Power BI Desktop, the Import option imports the selected tables and columns. The imported data can be used to create visuals.

The Import option contains the following process:
- While using the Get Data option, each selected table defines a query, which returns a set of data that can be edited and loaded into Power BI.
- Once the queries are loaded, the data related to queries is imported into the Power BI cache.
- Visuals created in Power BI Desktop query this imported data. Power BI is responsible for ensuring that the query is fast and efficient, and that changes to the visual are reflected instantly.
- Any update on data is not reflected in the visuals unless you refresh or reimport data.
- The report published to Power BI Service creates a dataset in Power BI Service. This dataset contains the imported data. You can set up scheduled data refresh, such as reimport data every day. You may also need to set up/configure an on-premise data gateway depending upon the location of the SQL Server.

DOI 10.1515/9781547400720-003

-   To ensure connectivity, the data imported into Power BI is automatically queried when you open an existing report or create a new one.
-   You can pin visuals as tiles to the dashboard. These tiles are refreshed automatically when you refresh the dataset.

Perform the following steps to use the Import option for importing database tables of SQL Server into Power BI Desktop:

1. Launch Power BI Desktop.
2. Click the **Get Data** button under the **External data** section of the **Home** tab. A list of data sources appears.
3. Select the **SQL Server** option from the list, as shown in Figure 3.1.

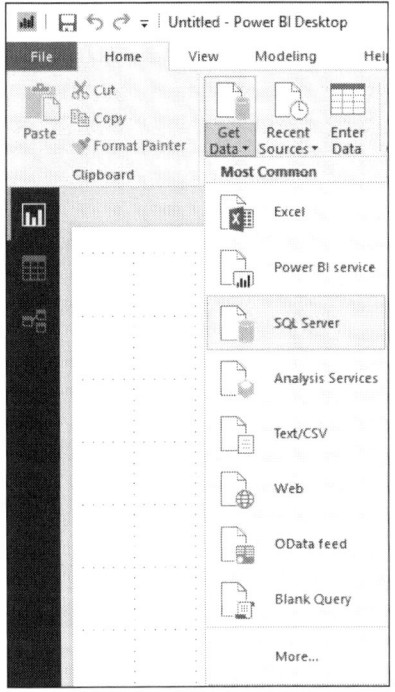

Figure 3.1: Selecting the SQL Server option

The **SQL Server database** dialog box appears.

4. Enter the server name in the **Server** text box.
5. Enter the name of the database in the **Database** text box.
6. Select the **Import** radio button under the **Data Connectivity mode** section.
7. Click the **OK** button, as shown in Figure 3.2.

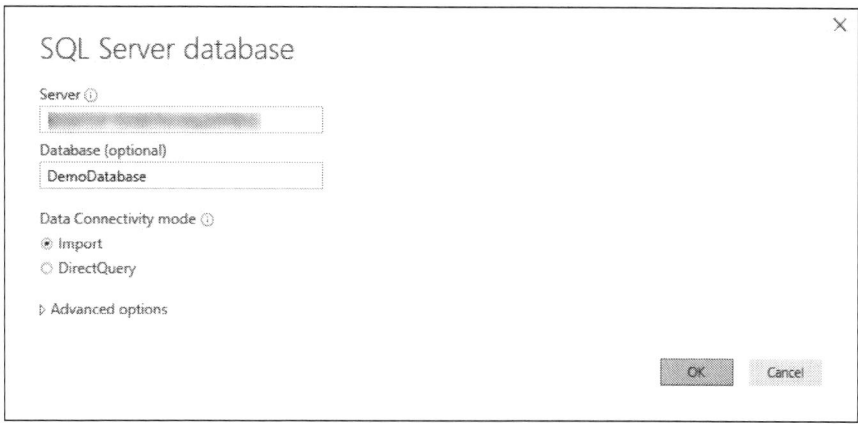

Figure 3.2: The SQL Server database dialog box

The **Navigator** window appears.

8. Select the checkboxes next to the table names appearing under the **Display Options** section in the left pane of the **Navigator** window. A preview of the table selected last appears in the right pane.

9. Click the **Load** button to load the tables into Power BI Desktop, as shown in Figure 3.3.

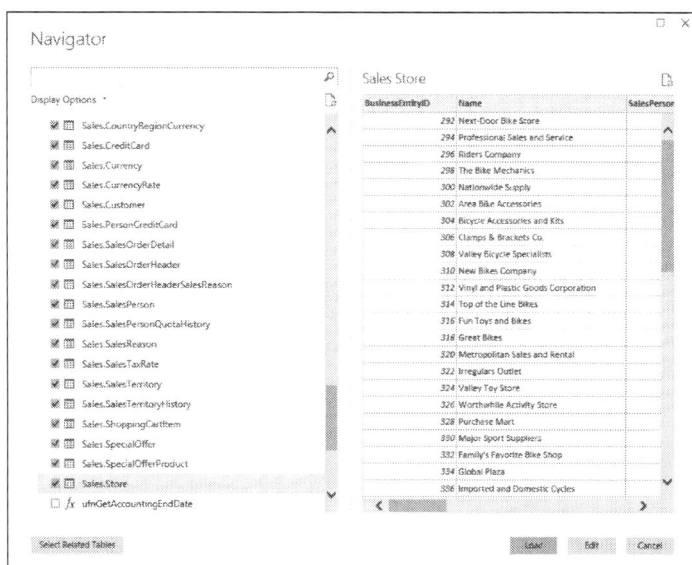

Figure 3.3: The Navigator window

The **Load** pane appears displaying the progress of each table being imported into Power BI Desktop, as shown in Figure 3.4.

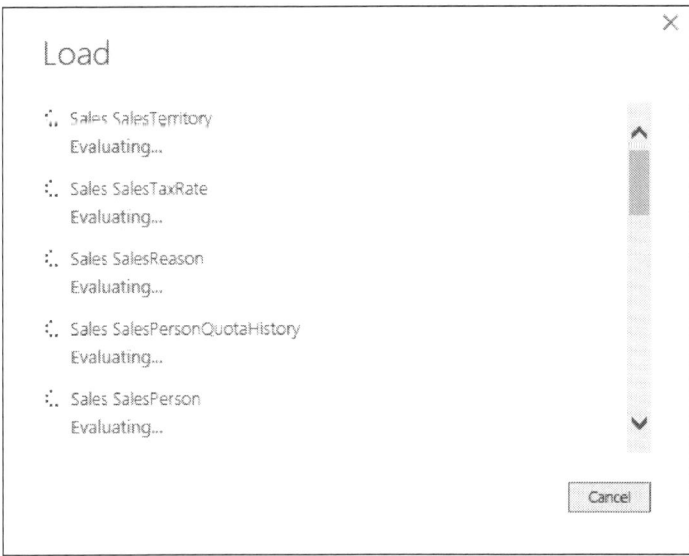

Figure 3.4: The Load pane

Once the loading process is done, the imported tables are displayed in the **FIELDS** pane, as shown in Figure 3.5.

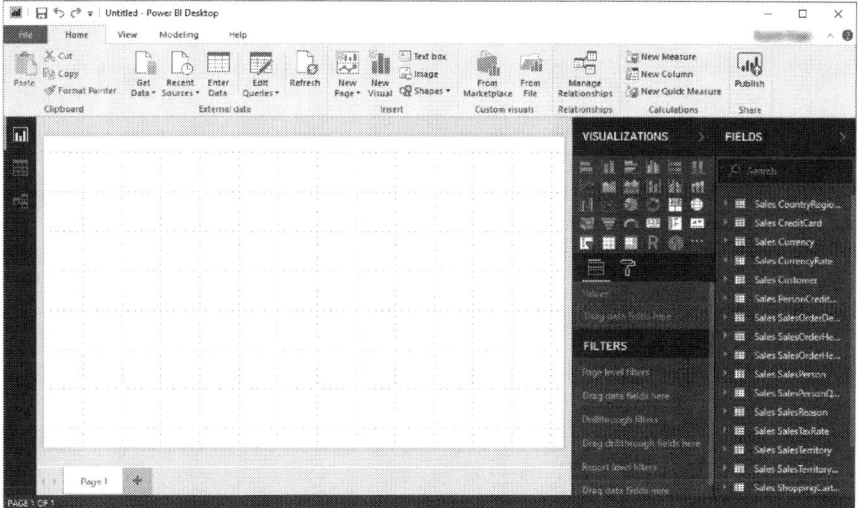

Figure 3.5: Imported tables

Using the DirectQuery Option

The DirectQuery option does not import or copy data to Power BI Desktop. However, the FIELDS pane displays the selected tables and columns. When you use the DirectQuery option, Power BI Desktop queries the connected data source each time you create a visual or report. This means that you get an updated visual each time. You can refresh the visuals with updated data by clicking the Refresh button under the Home tab.

Perform the following steps to use the DirectQuery option:

1. Launch Power BI Desktop.
2. Click the **Get Data** button under the **External data** section of the **Home** tab. A list of data sources appears.
3. Select the **SQL Server** option from the list. The **SQL Server database** dialog box appears.
4. Enter the server name in the **Server** text box.
5. Enter the name of the database in the **Database** text box.
6. Select the **DirectQuery** radio button under the **Data Connectivity mode** section.
7. Click the **Advanced options** button to view the advanced options. The associated options appear.

8. Enter the command timeout in the **Command timeout in minutes (optional)** text box. You can also leave this field blank as it is optional.
9. Enter the desired SQL statement in the **SQL statement (optional, requires database)** text area.
10. Select the **Include relationship columns** checkbox.
11. Click the **OK** button, as shown in Figure 3.6.

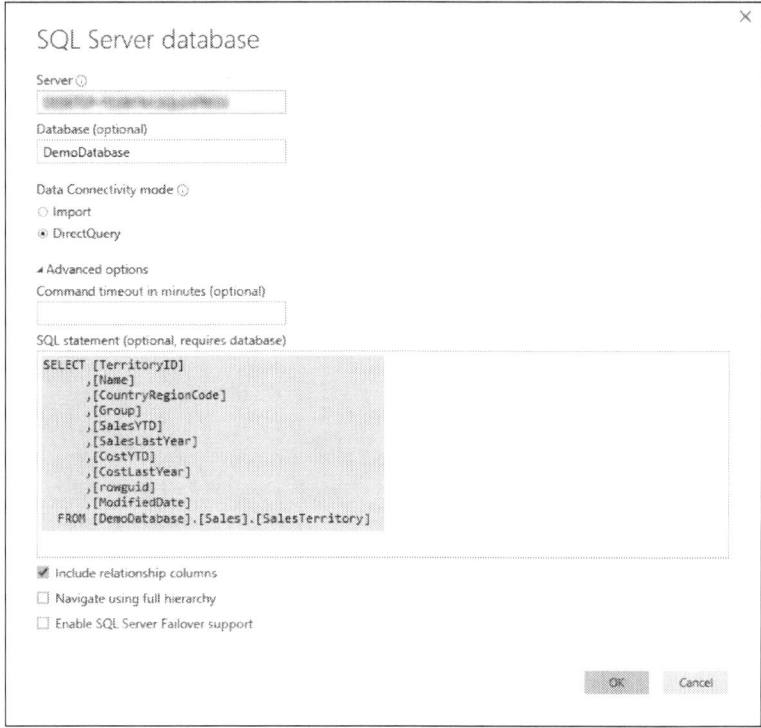

Figure 3.6: The SQL Server database dialog box

A window appears displaying the name of the database followed by the server name. It also displays the table view of the query made to the database.

12. Click the **Load** button to load the query, as shown in Figure 3.7.

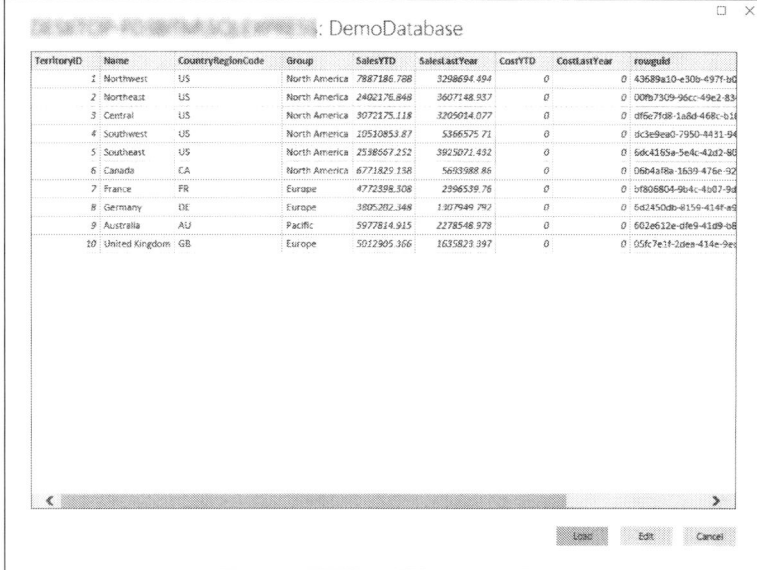

Figure 3.7: Loading a query

The **Create connections** message box appears displaying the progress of creating connections between Power BI and queried SQL Server statement, as shown in Figure 3.8.

Figure 3.8: The Create connections message box

Once the connection is established successfully, a query with the specified fields is added in the **FIELDS** pane, as shown in Figure 3.9.

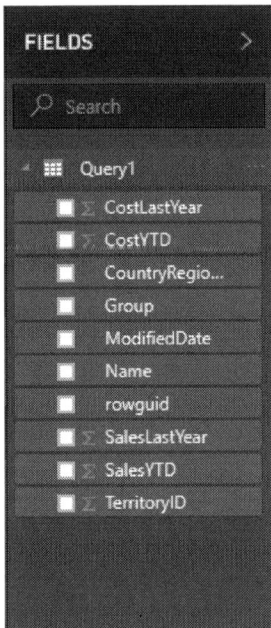

Figure 3.9: Displaying fields of a query

Advantages of Using the DirectQuery Option

There are several advantages of using the DirectQuery option. Some of them are as follows:

- It allows users to create visuals over very large datasets.
- The reports created by this option always contain the latest data.
- It supports large datasets, which means that it is not limited to 1 GB dataset.
- It supports several data modeling and data transformations.

Limitations of Using the DirectQuery Option

In addition to the advantages of the DirectQuery option, few limitations are also associated with it. Some of them are as follows:

- The tables selected for Power BI must be taken from a single database.
- The complex query of Query Editor may result in an error. This error can be rectified by using the Import option in place of the DirectQuery option.
- The relationship filtering can not be enabled in both the directions.
- It does not support time intelligence features, such as year, quarter, month, day, etc., of date columns.
- It supports a limited number of DAX expressions for creating a measure.

- It limits the number of returned data rows to 1 million, which in turn does not affect calculations used for creating the dataset.

Considerations When Using the DirectQuery Option

At the time of using the DirectQuery option to connect to SQL Server database tables in Power BI, you should consider the following three parameters:
1. Performance and load
2. Features
3. Security

Performance and Load

The performance of the DirectQuery option depends on the response time taken by the back-end source to respond to the query sent by the DirectQuery requests to the source database. The response time also affects the visuals refresh rate, which means that the visuals are refreshed as per the response time. The recommended response time for visuals to represent updated data is less than five seconds. The response time of thirty seconds is acceptable. After publishing the report to Power BI Service, the user receives an error if the query times out when it takes longer than a few minutes to respond.

The number of Power BI users consuming the published report is called the "Load on the database," which is greatly influenced by using Row Level Security (RLS). When we use RLS on a dashboard tile and refresh the tile, one query per user is generated for the database that in turn increases the load on the database and thus affects the performance.

Features

The DirectQuery option does not support all features in Power BI Desktop. Some features are supported with some limitations. Some features of Power BI Service are also not available when we use the DirectQuery option. For example, the datasets created using the DirectQuery option do not support the Quick Insights feature. Therefore, you should consider the limitations of such features to determine the use of the DirectQuery option.

Security

Once a report is published to Power BI Service, all users will be required to use credentials to connect to the back-end data source and consume the report. This condition is similar to the one used when data is imported through the Import option. The same data is available for all users irrespective of the security rules described in the back-end data source. The DirectQuery option is used for implementing per-user security for customers.

Data Modeling

Data modeling defines the way the data is modeled into Power BI. You do not need to put your data into one table. Instead, different tables can be imported from different data sources and a relationship can be defined between these tables to easily model the data. You can create calculated columns and apply configurations to view data segments in Power BI. You can also apply these configurations to visuals created in Power BI. We will cover the following items in this section:
- Creating relationships among tables
  - o Setting cardinality of relationships
  - o Understanding cross-filtering
- Using DAX
- Using calculated columns
- Using calculated tables

Creating Relationships Among Tables

As discussed earlier, you can import different tables from different data sources into Power BI. These tables may have a large amount of data. You may also need to do analysis using data from these tables. In this case, it will be hard to examine how these tables are related to each other. Therefore, you need to create relationships between these tables so that you can calculate results correctly and get the correct information in your reports.

In Power BI Desktop, you can easily create relationships between tables. You can either use the Autodetect feature to automatically create relationships among tables, or you can create it manually.

Note
You may need to make some edits to the relationship that was created automatically.

The **Relationships** view displays the relationship among tables, as shown in Figure 3.10.

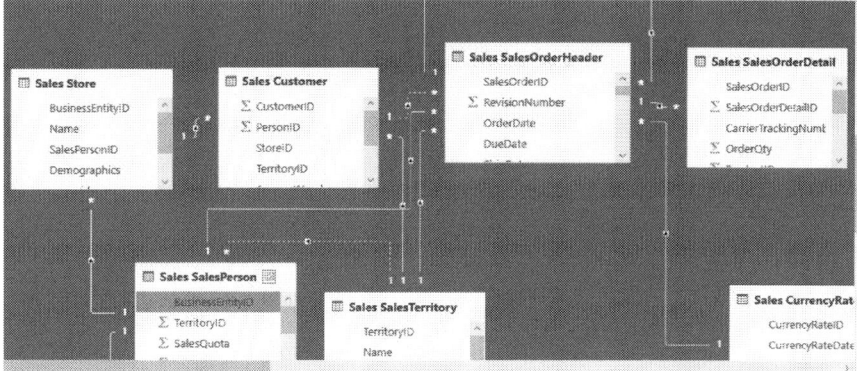

Figure 3.10:   The Relationships view

Using the Autodetect feature

Perform the following steps to create a relationship using the Autodetect feature:

1.   Click the **Manage Relationships** button under the **Relationships** section of the **Home** tab, as shown in Figure 3.11.

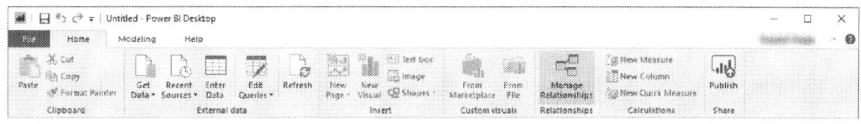

Figure 3.11:   Clicking the Manage Relationships button

The **Manage relationships** window appears.

2.   Click the **Autodetect** button to detect relationship automatically, as shown in Figure 3.12.

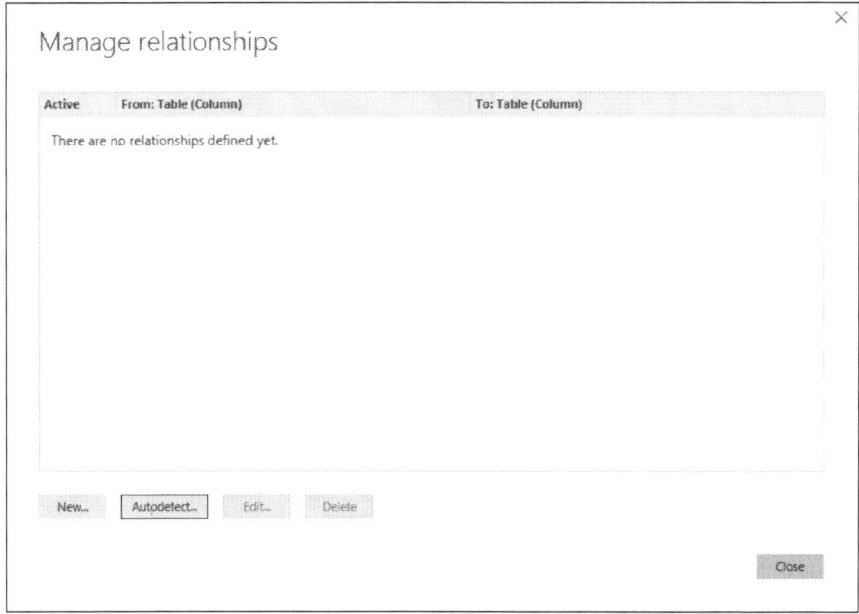

Figure 3.12: Using the Autodetect button

The **Detecting relationships** message box appears displaying the progress of the relationship detection, as shown in Figure 3.13.

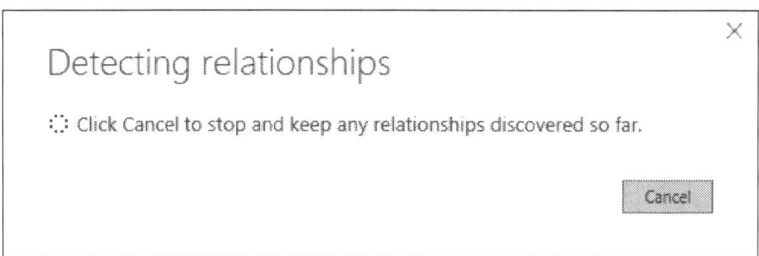

Figure 3.13: The Detecting relationships message box

The **Autodetect** message box appears once the detection is done.

3. Click the **Close** button to close the **Autodetect** message box, as shown in Figure 3.14.

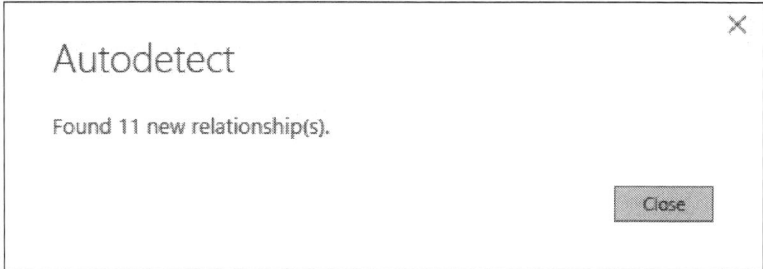

Figure 3.14: The Autodetect message box

The **Manage relationships** window displays the relationships among different tables, as shown in Figure 3.15.

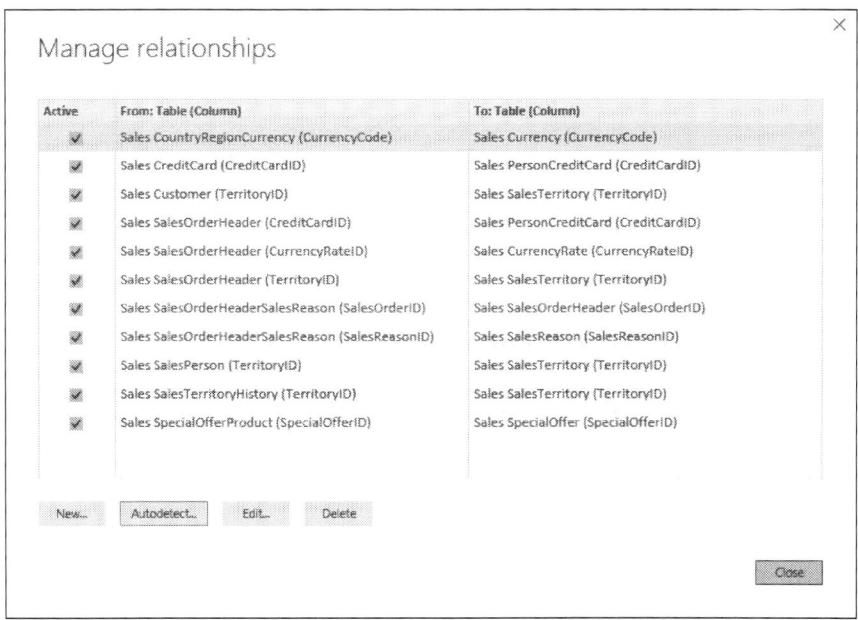

Figure 3.15: The Manage relationships window with relationships

4. Click the **Close** button to close the **Manage relationships** window.

You can see the relationships among tables in the **Relationships** view, as shown in Figure 3.16.

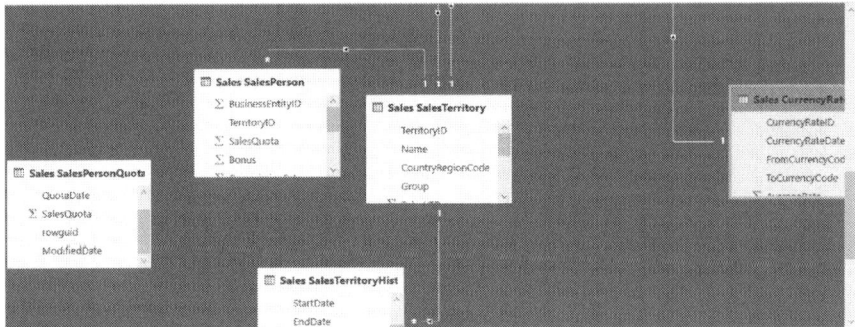

Figure 3.16: Viewing relationships in Relationships view

Creating Relationships Manually

Perform the following steps to create a relationship manually:

1. Click the **Manage Relationships** button under the **Relationships** section of the **Home** tab. The **Manage relationships** window appears.
2. Click the **New** button to create relationships manually.

The **Create relationship** dialog box appears.

3. Select the desired table from the first drop-down list. A list of columns associated with the selected table appears.
4. Select the desired column that you want to relate to another column.
5. Select another table from the second drop-down list. A list of columns associated with the selected table appears.
6. Select the desired column that you want to relate to the column selected in the first table.
7. Leave all other fields intact.
8. Click the **OK** button, as shown in Figure 3.17.

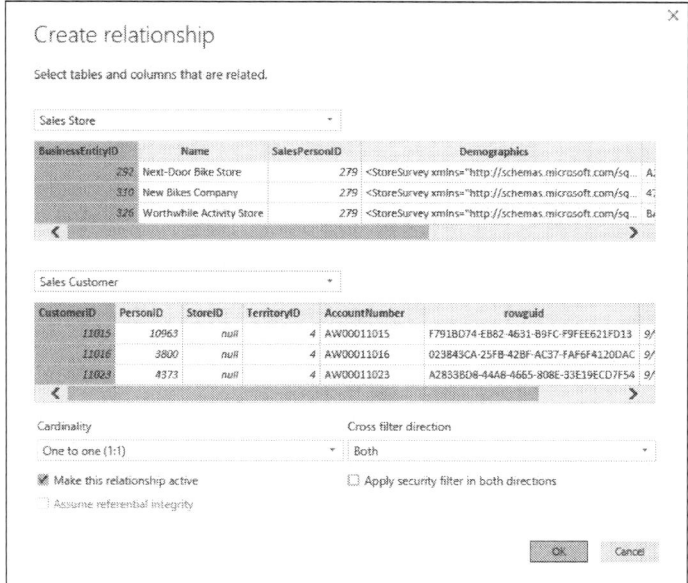

Figure 3.17: The Create relationship dialog box

The **Manage relationships** window appears and shows the new relationship.

9. Click the **Close** button to close the **Manage relationships** window, as shown in Figure 3.18.

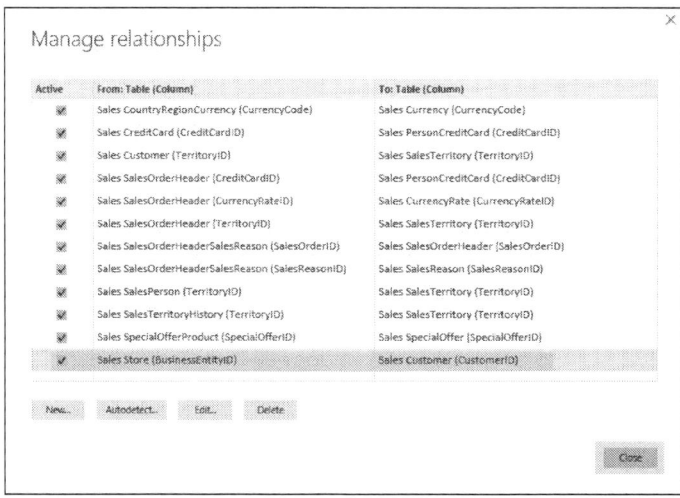

Figure 3.18: The Manage relationships window

10. Open the **Relationships** view to analyze the relationship between the selected tables, as shown in Figure 3.19.

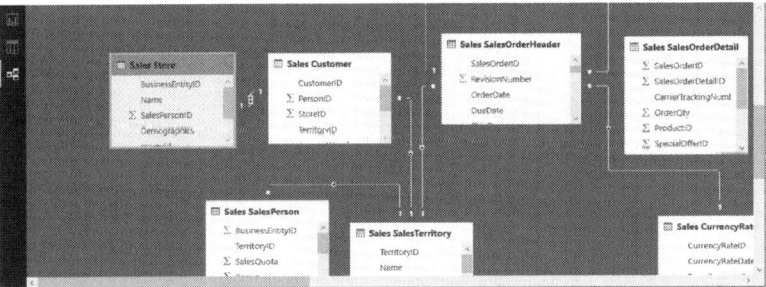

Figure 3.19: The Relationships view

Understanding Cardinality

Cardinality is defined as the degree of relationship, which specifies the number of occurrences of the first table linked to the number of occurrences of the second table. Power BI supports the following three types of cardinalities:

- **Many to One (*:1):** This is the default type of cardinality set in Power BI. It states that multiple occurrences in the first table are linked to a single occurrence in the other table, called the Lookup table.
- **One to One (1:1):** States that a single instance of a value in the first table is linked to only one instance of a specific value in the other table, called the Lookup table.
- **One to Many (1:*):** States that a single instance of a value in the first table is (or can be) linked to multiple occurrences in the other table.

Usually, the cardinality of the relationship is set automatically by Power BI Desktop. However, you may need to change it manually when you update data. The cardinality of the relationship can be set by using the Cardinality drop-down list. Let us demonstrate the concept of cardinality with an example.

We have two tables, named ProjectDetails and ProjectPriority. These tables contain the following data.

ProjectDetails

ProjectName	StartDate	EndDate
Project A	5/15/2017	10/20/2017
Project B	10/10/2017	12/12/2017

ProjectPriority

ProName	Priority
Project A	1
Project B	2
Project C	3
Project D	3

We can establish a relationship between these tables based on the ProjectName column in the ProjectDetails table and the ProName column in the ProjectPriority table. The cardinality between these tables can be set as One to One because we will not see any repeating values in the ProjectName column when we combine these tables. However, data update may occur over time, which will add more fields to the table. Now, the updated ProjectDetails table contains the following data:

ProjectName	StartDate	EndDate
Project A	5/15/2017	10/20/2017
Project B	10/10/2017	12/12/2017
Project A	10/21/2017	12/20/2017

When we combine both tables, we will get the following table:

ProjectName	Priority	StartDate	EndDate
Project A	1	5/15/2017	10/20/2017
Project B	2	10/10/2017	12/12/2017
Project C	3		
Project D	3		
Project A	1	10/21/2017	12/20/2017

From the above table, we can identify that the ProjectName column has repeating values, and One to One cardinality will not work. Therefore, we need to set Many to One cardinality in this case.

Cross-Filtering

Cross-filtering direction specifies the direction of the filter applied to related tables. Power BI provides the following two options for cross-filtering directions:

1. **Single:** When this option is applied to the relationship, filtering is applied to the table where values are being grouped. Note that the relationship will have a single direction when a Power Pivot is imported into Excel.
2. **Both:** This is the default direction for the related tables. It specifies that both the related tables act as a single table for filtering purposes. This cross-filtering option works well with a table surrounded by several lookup tables. Such configuration is known as Star schema configuration.

Using Data Analysis Expressions

Data Analysis Expressions (DAX) is an expression (a set of functions, constants, and operators) written to apply calculations on data available in your model. This newly created data can be used for creating visuals.

DAX formulas help to solve real business problems by allowing users to fetch the required information from the available data. This means that users can pull data from different tables according to their requirements and apply calculations on the data to view the required results. For example, your database could contain tables for sales, and for product descriptions. Querying for a product's sale can be an easy task. However, you may also need to calculate total sales of products that are sold in a region and within the specified time duration. Here, you need to use DAX to perform this task easily.

The fundamental concepts of DAX are as follows:
1. Syntax
2. Functions
3. Context

Syntax
Syntax defines the way a formula is written. A formula is made up of several elements, called syntax elements. Figure 3.20 shows a DAX formula.

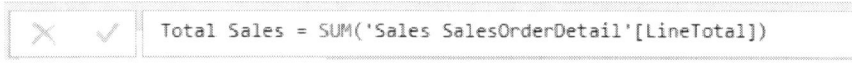

```
Total Sales = SUM('Sales SalesOrderDetail'[LineTotal])
```

Figure 3.20: DAX formula

A description of the syntax elements listed in above formula is as follows:
1. "Total Sales" is the name of measure.

2. The equals sign operator (=) states that the formula has started.
3. "SUM" is a DAX function, which returns the sum of all the numbers available in the specified column.
4. The parenthesis "()" encloses the expression having one or more arguments. There should be at least one argument for a function.
5. "Sale SalesOrderDetail" is the name of the referenced table.
6. "[LineTotal]" is the name of the column in the "Sale SalesOrderDetail" table.

Note

The easiest way to understand a DAX formula is separating the syntax elements into a natural language. For example, the formula listed in Figure 3.20 can be read as:

To get "Total Sales," calculate (=) the SUM of values in the [LineTotal] column of the "Sale SalesOrderDetail" table.

Functions

A function is a predefined formula that returns a value in a specified order by performing calculations on specific values, known as arguments. An argument can be a formula, function, number, or text. Power BI supports a variety of functions. Table 3.1 lists some useful functions with their syntaxes and descriptions.

Table 3.1: Commonly Used Functions

Function Name	Syntax	Description
CALENDAR	CALENDAR(<start_date>, <end_date>)	Returns a continuous set of dates in a single column within the specified range.
DATE	DATE(<year>, <month>, <day>)	Returns the specified date.
DATEDIFF	DATEDIFF(<start_date>, <end_date>, <interval>)	Returns the number of intervals between the specified dates.
DATEVALUE	DATEVALUE(date_text)	Returns a datetime value based on the locale settings of your computer.
DAY	DAY(date)	Returns the day of the month in integer format.
NOW	NOW()	Returns the current date and time.
TODAY	TODAY()	Returns the current date with time set to 12:00:00 PM.
CALCULATE	CALCULATE(<expression>,<filter1>,<filter2>...)	Returns an expression based on the specified filters.
AND	AND(<logical1>,<logical2>)	Returns TRUE only when both the arguments are TRUE, otherwise returns FALSE.

Table 3.1. (continued)

Function Name	Syntax	Description
IF	IF(<logical_test>,<value_if_true>, value_if_false)	Returns first value when the specified expression (<logical_test>) is TRUE, otherwise returns second value.
OR	OR(<logical1>,<logical2>)	Returns TRUE when one of the arguments is TRUE and returns FALSE when both the arguments are FALSE.
SUM	SUM(<column>)	Returns the sum of all the values specified in a column.
TRUNC	TRUNC(<number>, <num_digits>)	Returns a truncated number as an integer by eliminating the decimals or fractional part of the number.

Note
A complete list of functions is available at the following link:
https://msdn.microsoft.com/en-us/library/ee634396.aspx

Context

While discussing DAX, it is advisable to understand context. DAX supports the following two types of contexts:

1. **Row context:** This type of context is applicable when a function is available within a formula and this function can examine a single row in a table through filters. This function inherits and applies row context for each filtering rows of the table. This context is usually applied to measures.
2. **Filter context:** This type of context deals with applying multiple filters in a calculation that defines a result or value. They do not override row context, but are used with row context. Most reports use filter context. For example, when you apply a visual to a field named TotalSales, and add filters like Year and Region to it, it means that you are applying a filter context, which provides a data subset based on the specified year and region.

Using Calculated Columns

You can add custom columns to the tables available in your model. After adding these columns, you can add data to these columns either by loading values from the data source or by creating a DAX formula. You can create a calculated column by clicking the New Column button.

Calculated columns that you create in Power BI appear in the FIELDS pane similar to other columns already available under the respective table. You can also assign the required name for the calculated column and use this field for creating visuals like other fields.

Perform the following steps to create a calculated column:
1. Click the **New Column** button under the **Calculations** section of the **Modeling** tab. A new column is added to the selected table in the **FIELDS** pane and the formula bar appears above the Report canvas, as shown in Figure 3.21.

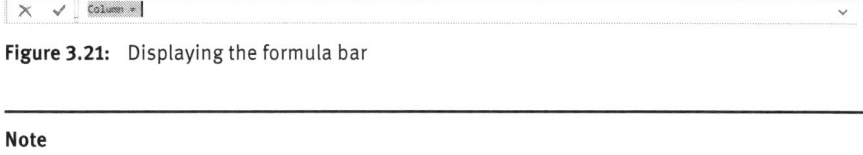

Figure 3.21: Displaying the formula bar

Note
The formula bar is the area where you can rename the column and specify a DAX formula.

---

2. Enter the desired DAX formula in the formula bar. In our case, we are concatenating values of two columns into one, as shown in Figure 3.22.

Figure 3.22: Using DAX formula

From the above figure, following considerations can be made:
   a. Name is the name of the calculated column.
   b. CONCATENATE is the name of the function that concatenates two strings.
   c. Sales_By_Region[FirstName] states the FirstName column in the Sales_By_Region table.
   d. Sales_By_Region[LastName] states the LastName column in the Sales_By_Region table.
3. Click the **OK** (✓) icon to accept the changes.

After you click the **OK** (✓) icon, the calculated column is created in the specified table. You can view a visual for the calculated column by selecting the desired visual from the VISUALIZATIONS pane and selecting fields to display in the visual from the FIELDS pane, as shown in Figure 3.23.

Figure 3.23: Creating a visual for the calculated column

Using Calculated Tables

The concept of creating a calculated table is similar to creating a calculated column. You can create a calculated table in Power BI Desktop by simply using the New Table button under the Calculations section of the Modeling tab. You can add data to the columns of a calculated table either manually or by using the DAX formula. You can specify the desired name for the calculated table as per your requirements.

The concept of calculated tables can be better understood with the help of an example. In our example, a company has two offices located in the Northwest and the Southwest regions. Each office maintains a list of its employees. The company director wants to see a list of all its employees working in both the offices. Doing this work manually can be a tricky process as the number of employees grows. In Power BI, the calculated table can be used such that it joins both the tables and lists all the employees in a single table.

Perform the following steps to create a calculated table:

1. Click the **New Table** button under the **Calculations** section of the **Modeling** tab. A formula bar appears, as shown in Figure 3.24.

Figure 3.24: A formula bar

2. Enter a DAX formula for joining two tables in the formula bar, as shown in Figure 3.25.

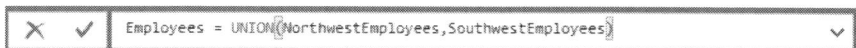

Figure 3.25: Entering a DAX Formula

From the DAX formula mentioned in the figure above, the following considerations can be made:

    a. "Employees" is the name of the calculated table.

    b. "=" is the operator.

    c. UNION is the name of the function that joins two tables named NorthwestEmployees and SouthwestEmployees.

3. Click the **OK** (✔) icon to save the DAX formula. A table named Employees is created and displayed similar to other tables in the FIELDS pane, as shown in Figure 3.26.

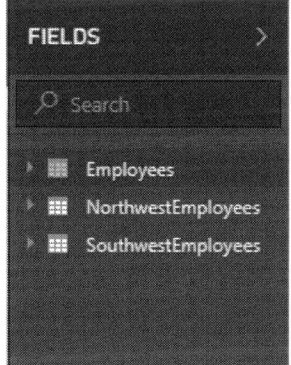

Figure 3.26: The FIELDS pane

You can view the content of this table in the Data view, as shown in Figure 3.27.

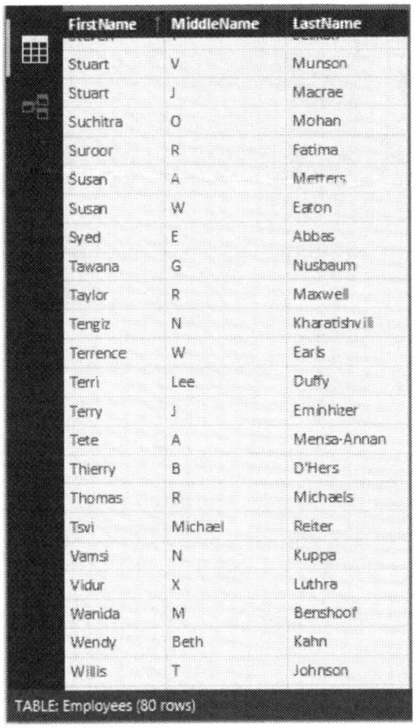

FirstName	MiddleName	LastName
Stuart	V	Munson
Stuart	J	Macrae
Suchitra	O	Mohan
Suroor	R	Fatima
Susan	A	Metters
Susan	W	Eaton
Syed	E	Abbas
Tawana	G	Nusbaum
Taylor	R	Maxwell
Tengiz	N	Kharatishvili
Terrence	W	Earls
Terri	Lee	Duffy
Terry	J	Eminhizer
Tete	A	Mensa-Annan
Thierry	B	D'Hers
Thomas	R	Michaels
Tsvi	Michael	Reiter
Vamsi	N	Kuppa
Vidur	X	Luthra
Wanida	M	Benshoof
Wendy	Beth	Kahn
Willis	T	Johnson

TABLE: Employees (80 rows)

Figure 3.27: Viewing the content of the table

Creating a Report

Once you are done with the process of pulling data from SQL Server and applying data modeling practices, you can create a report containing visuals in Power BI Desktop. As discussed earlier, the following two options can be used to get data from SQL Server to Power BI Desktop:

1. Using the DirectQuery option
2. Using the Import option

In this section, you will learn to create isolated reports using both options.

Creating a Report Using the DirectQuery Option

Perform the following steps to load data into Power BI Desktop using the DirectQuery option and create a report based on that data:

1. Launch Power BI Desktop.
2. Click the **Get Data** button under the **External data** section of the **Home** tab. A list of data sources appears.
3. Select the **SQL Server** option from the list. The **SQL Server database** dialog box displays.
4. Enter the server name in the **Server** text box.
5. Enter the name of the database in the **Database** text box.
6. Select the **DirectQuery** radio button under the **Data Connectivity mode** section.
7. Click the **Advanced options** button to view the advanced options. The associated options appear.
8. Enter the command timeout in the **Command timeout in minutes** text box. You can also leave this field blank since it is optional.
9. Enter the desired SQL statement in the **SQL statement (optional, requires database)** text area. In our case, we have entered the following SQL statement:

```
SELECT soh.SalesOrderID, soh.TotalDue,soh.SubTotal, soh.OrderDate,  c.CustomerID,
p.FirstName, p.LastName, RTRIM(sp.StateProvinceCode) as StateProvinceCode,
ad.City,  sp.Name as State, ctr.Name as Country, ad.PostalCode FROM Sales.
SalesOrderHeader AS soh INNER JOIN Sales.Customer AS c ON soh.CustomerID =
c.CustomerID INNER JOIN

Person.BusinessEntity AS b ON b.BusinessEntityID = c.PersonID INNER JOIN

Person.Person AS p ON p.BusinessEntityID = b.BusinessEntityID INNER JOIN

Person.BusinessEntityAddress AS a ON a.BusinessEntityID = b.BusinessEntityID INNER
JOIN

Person.Address AS ad ON ad.AddressID = a.AddressID INNER JOIN

Person.AddressType AS at ON at.AddressTypeID = a.AddressTypeID INNER JOIN

Person.StateProvince AS sp ON sp.StateProvinceID = ad.StateProvinceID INNER JOIN

Person.CountryRegion ctr ON sp.CountryRegionCode = ctr.CountryRegionCode

WHERE(c.PersonID IS NOT NULL)

    AND (at.Name = N'Home')

    AND (sp.CountryRegionCode = N'US')
```

10. Click the **OK** button, as shown in Figure 3.28.

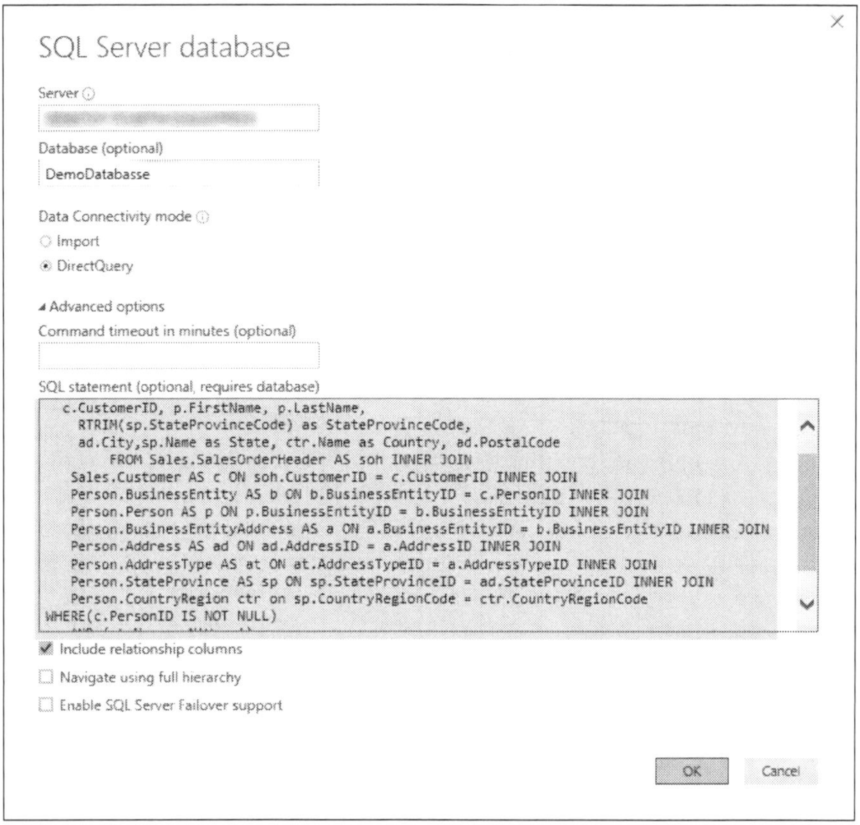

SQL Server database

Server ⓘ

[   ]

Database (optional)

DemoDatabasse

Data Connectivity mode ⓘ
○ Import
◉ DirectQuery

◢ Advanced options

Command timeout in minutes (optional)

[   ]

SQL statement (optional, requires database)

```
   c.CustomerID, p.FirstName, p.LastName,
    RTRIM(sp.StateProvinceCode) as StateProvinceCode,
    ad.City,sp.Name as State, ctr.Name as Country, ad.PostalCode
       FROM Sales.SalesOrderHeader AS soh INNER JOIN
  Sales.Customer AS c ON soh.CustomerID = c.CustomerID INNER JOIN
  Person.BusinessEntity AS b ON b.BusinessEntityID = c.PersonID INNER JOIN
  Person.Person AS p ON p.BusinessEntityID = b.BusinessEntityID INNER JOIN
  Person.BusinessEntityAddress AS a ON a.BusinessEntityID = b.BusinessEntityID INNER JOIN
  Person.Address AS ad ON ad.AddressID = a.AddressID INNER JOIN
  Person.AddressType AS at ON at.AddressTypeID = a.AddressTypeID INNER JOIN
  Person.StateProvince AS sp ON sp.StateProvinceID = ad.StateProvinceID INNER JOIN
  Person.CountryRegion ctr on sp.CountryRegionCode = ctr.CountryRegionCode
WHERE(c.PersonID IS NOT NULL)
```

☑ Include relationship columns

☐ Navigate using full hierarchy

☐ Enable SQL Server Failover support

[OK]   Cancel

Figure 3.28: The SQL Server database dialog box

A window appears displaying the name of the database followed by the server name. It also shows the table view of the query made to the database.

11. Click the **Load** button to load the query, as shown in Figure 3.29.

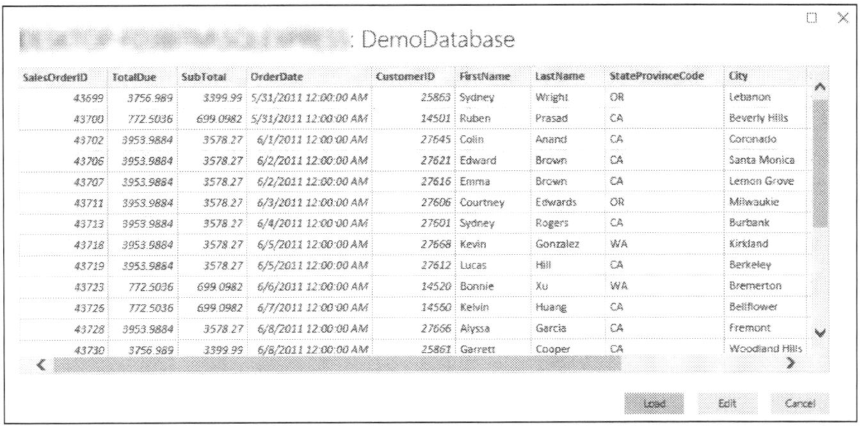

Figure 3.29: Loading a query

The **Create connections** message box appears displaying the process of creating connections between Power BI and the queried SQL Server statement. Once the connection is established successfully, a query (Query1) with the specified fields is added in the **FIELDS** pane.

12. Right-click the query. A context menu opens.
13. Select the **Rename** option from the context menu to rename it, as shown in Figure 3.30.

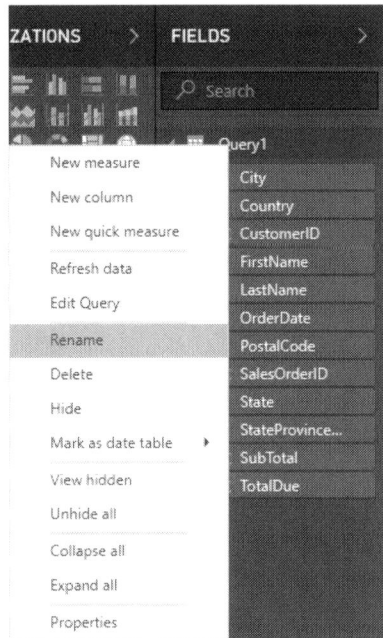

Figure 3.30: Renaming query

14. Replace the word **Query1** with the desired name. In our case, we have entered **Sales_By_Region**.
15. Click the **New Column** button under the **Calculations** section of the **Modeling** tab. The formula bar appears.
16. Type the following DAX expression to create a calculated column called **Name** that concatenates two columns called **FirstName** and **LastName**:

```
Name = CONCATENATE(Sales_By_Region[FirstName], Sales_By_Region[LastName])
```

17. Click the **OK** (✓) icon to save the DAX expression. This creates a calculated column titled **Name** under the **Sales_By_Region** table, as shown in Figure 3.31.

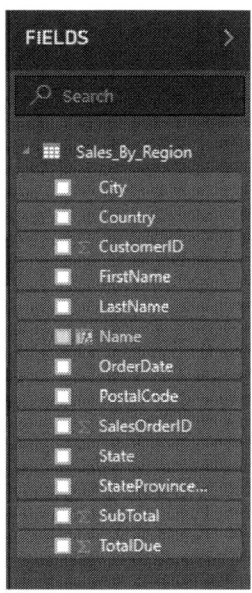

Figure 3.31: Creating a calculated column

18. Insert the desired visuals that you want to showcase in your report. In our case, we have created a report with a variety of visuals, as shown in Figure 3.32.

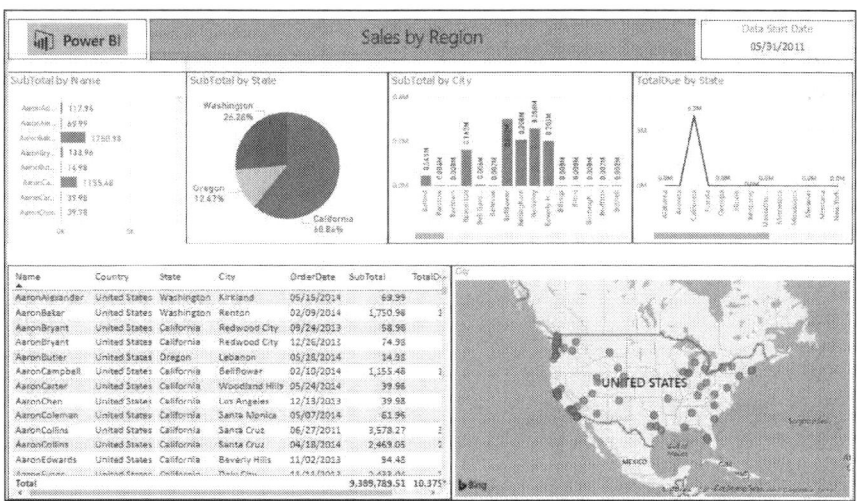

Figure 3.32: Creating a report

In the above report, we have performed the following tasks:

1. Inserted an image by using the Image button under the Insert section of the Home tab. Once inserted, we set its border to black in the FORMAT IMAGE pane.

2. Inserted a text box by clicking the Text box button under the Insert section of the Home tab and setting its font size. We have also set its background color.

3. Inserted a card visual by clicking the Card visual under the VISUALIZATIONS pane. We dragged the OrderDate field under the FIELDS pane to the Values field under the Fields section. Then, we added a title to the card. We also set the Category label to Off under the Format section. The border color is specified as black.

4. Added a clustered bar chart by clicking the Clustered bar chart icon under the VISUALIZATIONS pane. In the Fields section, we dragged Name to the Axis field and SubTotal to the Value field. The border color is specified as black.

5. Added a pie chart by clicking the Pie chart icon under the VISUALIZATIONS pane. In the Fields section, we dragged State to the Legend field and SubTotal to the Values field. The border color is specified as black. We have also selected the Percentage of total option for Label style under the Detail labels section.

6. Added a clustered column chart by clicking the Clustered column chart icon under the VISUALIZATIONS pane. In the Fields section, we dragged City to the Axis field and SubTotal to the Value field. The border color is specified as black. The Orientation value is set to Vertical under the Data labels section.

7. Added a line chart by clicking the Line chart icon under the VISUALIZATIONS pane. In the Fields section, we dragged State to the Axis field and TotalDue to the Values field. The border color is specified as black. A color is set for the TotalDue value under the Data colors section.

8. Added a table by clicking the Table icon under the VISUALIZATIONS pane. In the Fields section, we dragged Name, Country, State, City, OrderDate, SubTotal, and TotalDue to the Values field. The border color is specified as black. A color is set for the TotalDue value under the Data colors section. The Font family and Text size options under the Values section are set to Calibri and 12, respectively.

9. Added a map by clicking the Map icon under the VISUALIZATIONS pane. In the Fields section, we dragged City to the Location field. The border color is specified as black.

Creating a Report Using the Import Option

Perform the following steps to load the data into Power BI Desktop using the Import option and creating a report based on that data:

1. Launch Power BI Desktop.
2. Click the **Get Data** button under the **External data** section of the **Home** tab. A list of data sources appears.
3. Select the **SQL Server** option from the list. The **SQL Server database** dialog box appears.
4. Enter the server name in the **Server** text box.
5. Enter the name of the database in the **Database** text box.
6. Select the **Import** radio button under the **Data Connectivity mode** section.
7. Click the **OK** button. The **Navigator** window appears.
8. Select the desired tables from the **Navigator** window.
9. Click the **Load** button, as shown in Figure 3.33.

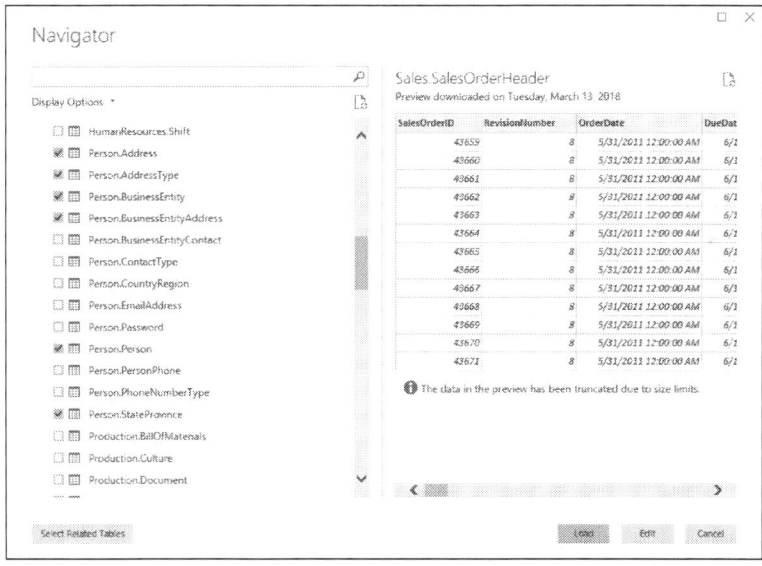

Figure 3.33: Loading tables

The **Load** dialog box appears displaying the progress of loading the selected tables. Once the selected tables are loaded, they are displayed in the **FIELDS** pane.

10. Insert the desired visuals that you want to showcase in your report. In our case, we have created a report showing different visuals, as shown in Figure 3.34.

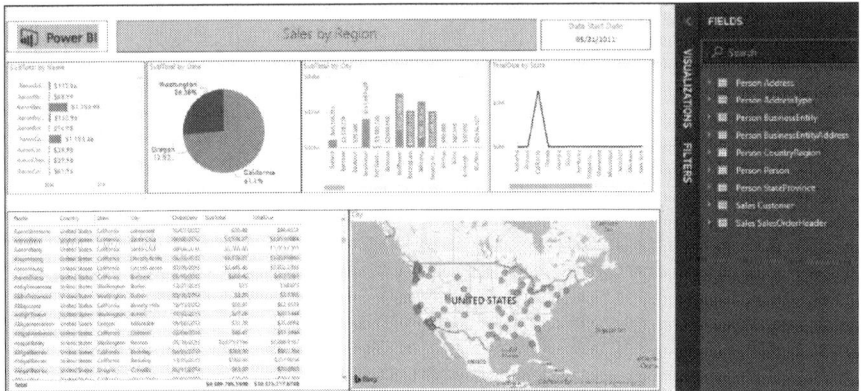

Figure 3.34: Creating a report

Saving the Report

Once you are done with the process of creating a report, you need to save it to an appropriate location. You should save the report regularly to avoid any loss of your work on the report due to an unexpected event like a power failure.

Perform the following steps to save a report:

1. Select the **File** tab on the Ribbon. The Backstage View appears.
2. Select the **Save As** option. The **Save As** dialog box appears.
3. Navigate to the location wherein you want to save your report.
4. Enter the desired name for the report in the **File name** combo box.
5. Click the **Save** button, as shown in Figure 3.35.

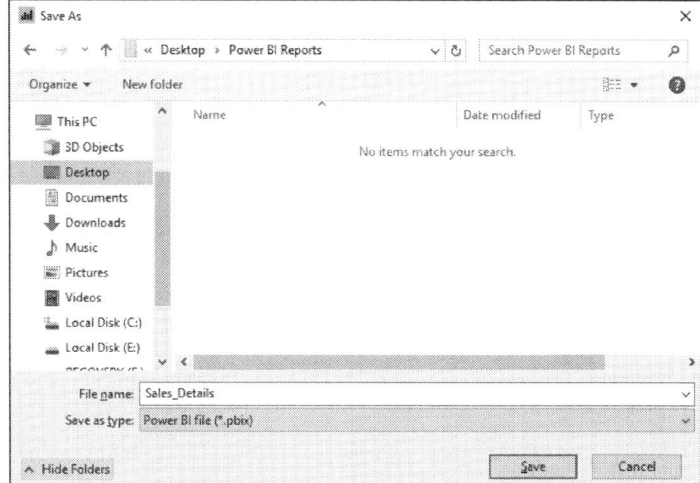

Figure 3.35: Saving a report

Once you save the report, the specified name for the report appears in the title bar, as shown in Figure 3.36.

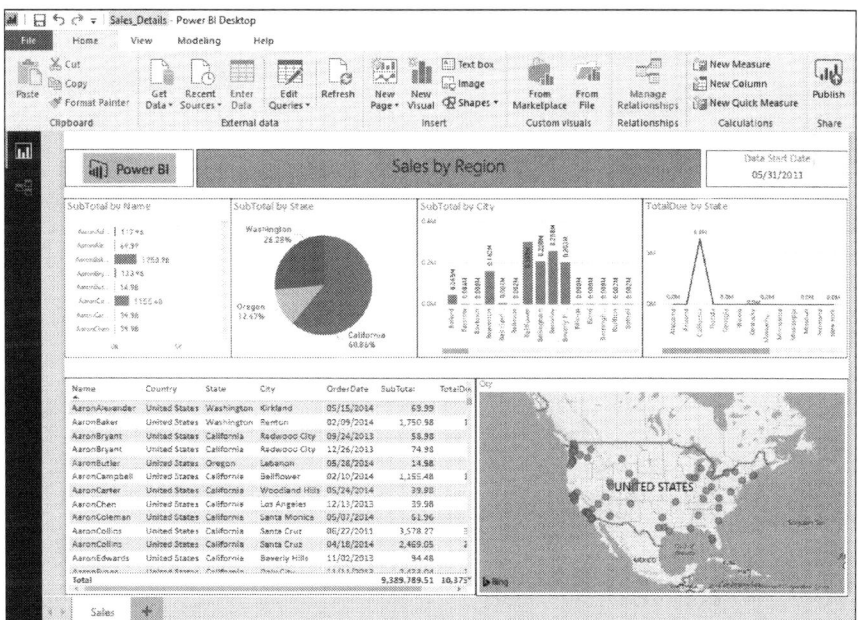

Figure 3.36: Displaying the title of the report

Publishing the Report

After designing a report in Power BI Desktop, you need to publish it to Power BI Service to make it available for others. When you publish a report to Power BI Service, it automatically creates a dataset into Power BI Service.

Perform the following steps to publish the report:

1. Click the **Publish** button under the **Share** section of the **Home** tab, as shown in Figure 3.37.

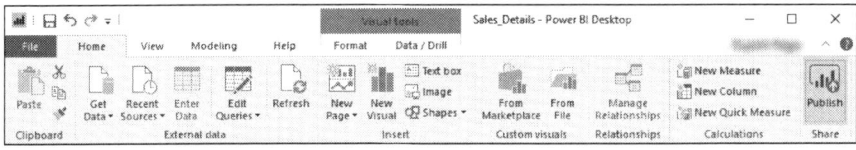

Figure 3.37: Clicking the Publish button

The **Publish to Power BI** dialog box appears.

2. Select the workspace from the **Select a destination** list box.
3. Click the **Select** button, as shown in Figure 3.38.

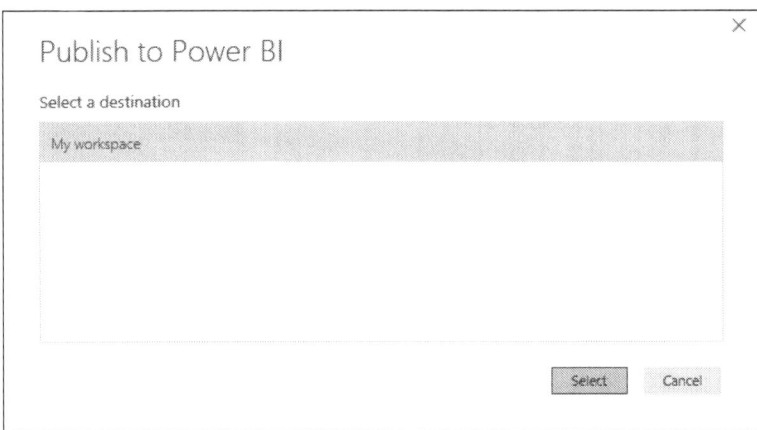

Figure 3.38: The Publish to Power BI dialog box

The **Publishing to Power BI** message box appears displaying the progress bar of publishing the report to Power BI, as shown in Figure 3.39.

Figure 3.39: The Publishing to Power BI message box

Once the publishing is complete, you will receive the message "Publishing succeeded." However, you will also receive a warning, stating that the published report can not connect to the data source due to the unavailability of a gateway.

4. Click the **Got it** button to close the message box, as shown in Figure 3.40.

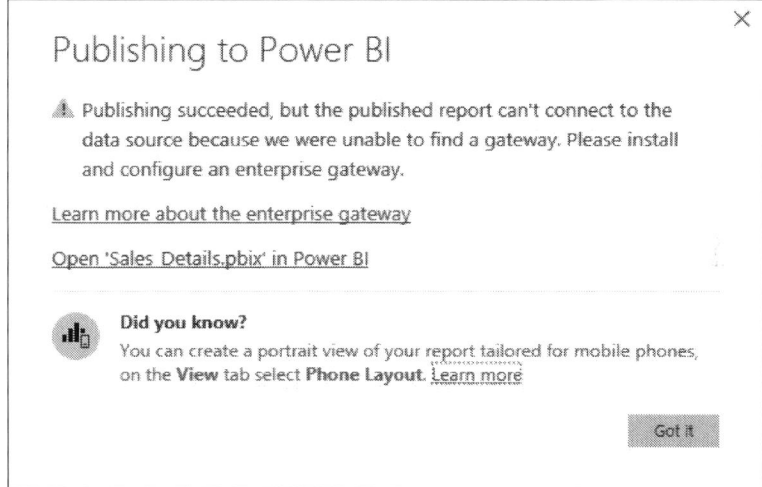

Figure 3.40: Publishing succeeded message box

Note
We will learn about installing and configuring gateway in the next section.

Gateway Setup

As discussed in the previous section, you need to install and configure an enterprise gateway to connect the published report to the data source. This section covers the following topics:
1. Understanding gateways
2. Types of gateways
3. Downloading and installing the on-premises data gateway
4. Configuring gateway
5. Adding a data source

Understanding Gateways

A gateway is software that allows users to access data located on an on-premises system or network so that it can be used in a cloud service later. Only authorized users are permitted to access the gateway. Similar to a gatekeeper who allows only authorized personnel to enter through the gate, the gateway attends all connection requests but grants access only to those users who meet certain criteria. Instead of using the entire database for creating reports and dashboards in Power BI, a subset of data can be used and the entire database can be placed over the on-premises network. Additional functions of a gateway include:
– Encrypting the data passing through it
– Compressing the data for secure access
– Using passwords for establishing a connection to data sources

Types of Gateways

Power BI provides two types of gateways, both of which work in the same manner. These gateways are listed below:
1. On-premises data gateway (personal mode)
2. On-premises data gateway

On-Premises Data Gateway (Personal Mode)
On-premises data gateway (personal mode) authenticates only one person to connect to data sources. It does not allow sharing of reports. It can only be used by Power BI. This gateway is for personal use wherein the user installs the gateway on his/her computer and the data source is located on-premises.

This type of gateway is ideal for situations as in the following example. If you have a workbook that contains sales data for several years and you want to create a Power BI dashboard that contains tiles displaying sales figures based on different parameters. You are the owner of the report and use the same dataset for creating Power BI reports, you need an on-premises data gateway (personal mode).

Did You Know?
The on-premises data gateway (personal mode) is an updated version of the earlier personal gateway called Power BI Gateway – Personal.

On-Premises Data Gateway

The on-premises data gateway can be used and shared by multiple users. This type of gateway can be used by multiple services including Power BI, PowerApps, and Azure Logic Apps, etc. It supports scheduled refresh as well as DirectQuery for Power BI.

This setup is ideal for situations where you are working for an organization that accesses data from different data sources including Analysis Services, SAP, Oracle, and IBM, etc., and several people in the organization want to access the database to create multitudes of reports. To gain access to these sources, you need the on-premises data gateway to share the reports with several people in the organization.

The following are some considerations while using on-premises data gateway:
1. You can not install a gateway on a domain controller.
2. You should not install a gateway on a laptop that could experience broken internet connectivity.
3. You should avoid installing gateway on a system running on the wireless network as it might affect its performance.

Downloading and Installing the On-Premises Data Gateway

After publishing a report in Power BI Service, you can access your report at the https://app.powerbi.com link with your credentials. As you open the published report, you will receive an error, as shown in Figure 3.41.

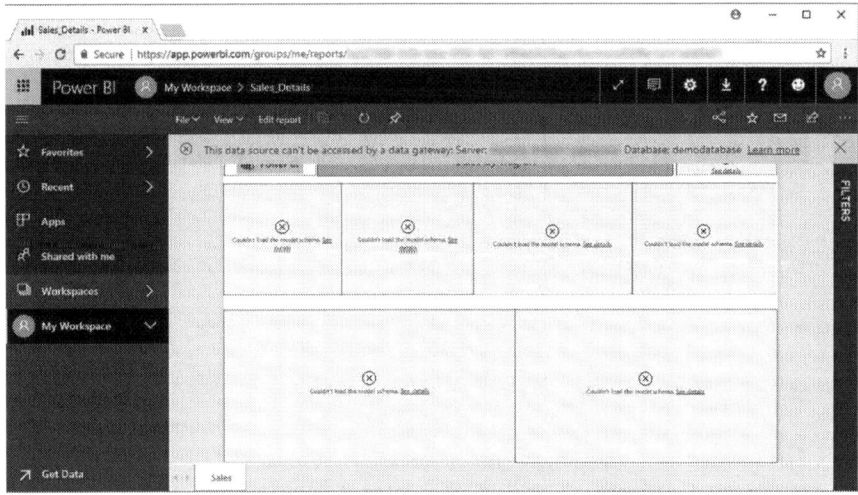

Figure 3.41: Displaying the error

From the above figure, you will see that visuals are not appearing in the report. This is due to the inaccessibility of data gateway to reach the data source. This problem can be rectified by installing an on-premises data gateway on the machine running SQL Server and configuring it such that it can be used by Power BI.

Perform the following steps to download and install an on-premises data gateway:
1. Browse the following link:
   https://app.powerbi.com/
2. Sign in to Power BI Service with the same credentials used in Power BI Desktop.
3. Click the **Download** icon button. A list of options appears.
4. Click the **Data Gateway** option from the list of options, as shown in Figure 3.42.

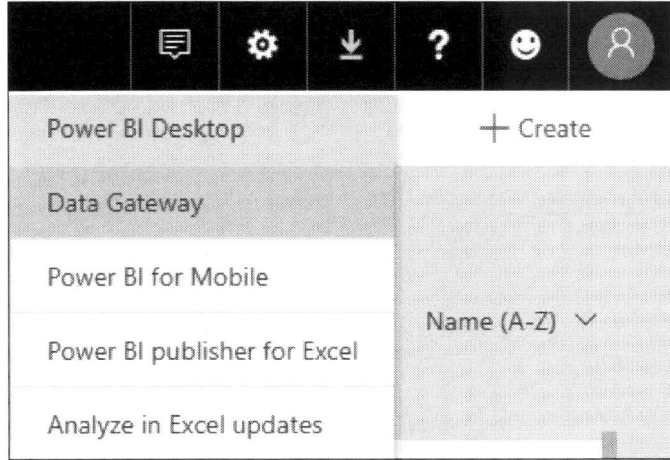

Figure 3.42: Clicking the Data Gateway option

You are redirected to a link to download gateway.

5. Click the **DOWNLOAD GATEWAY** button, as shown in Figure 3.43.

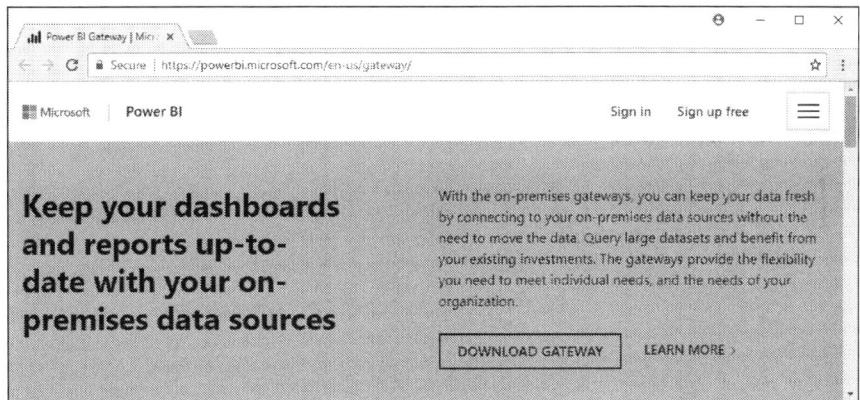

Figure 3.43: Download page

Once the download is complete, you will see the **PowerBIGatewayInstaller.exe** file in the **Downloads** folder.

6. Double-click the **PowerBIGatewayInstaller.exe** in the **Downloads** folder. The **On-premises data gateway installer** wizard appears.

7. Click the **Next** button to start data gateway installation, as shown in Figure 3.44.

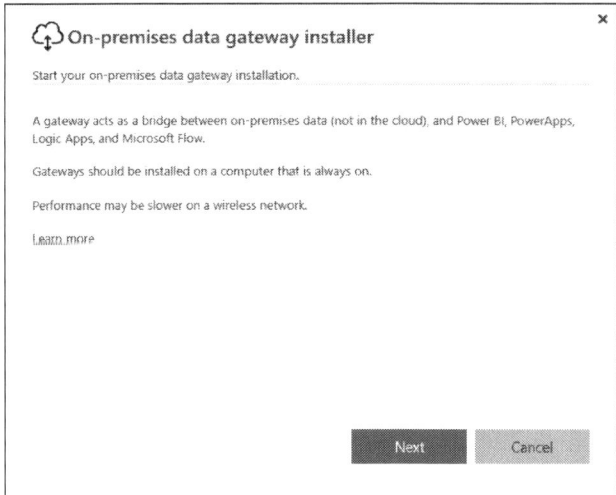

Figure 3.44: Starting the installation

The **Choose the type of gateway you need** page of the **On-premises data gateway installer** wizard appears.

8. Select the **On-premises data gateway (recommended)** radio button to install the on-premises data gateway.

9. Click the **Next** button, as shown in Figure 3.45.

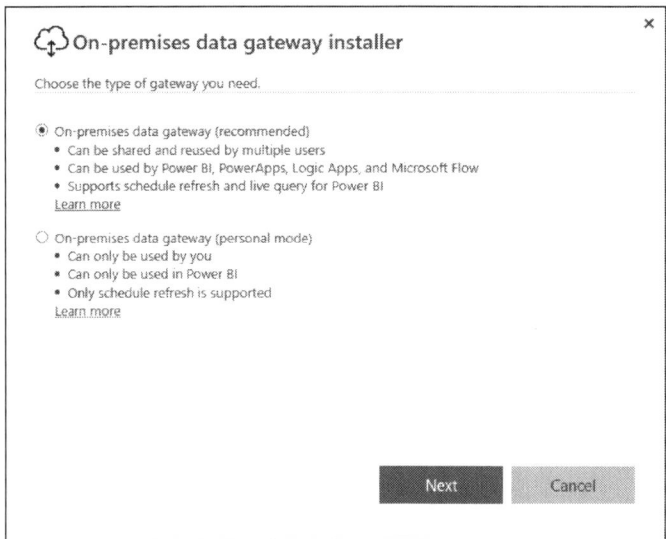

Figure 3.45: Selecting the gateway type

The **Getting ready to install the on-premises data gateway** page appears, displaying the progress of the installation, as shown in Figure 3.46.

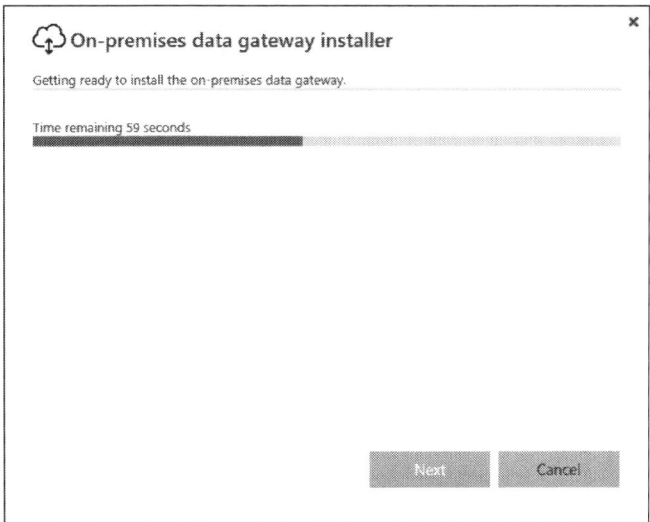

Figure 3.46: The Getting ready to install the on-premises data gateway page

The **Reminder before you install** page appears. This page provides a warning and reminder about using the gateway.

10. Click the **Next** button, as shown in Figure 3.47.

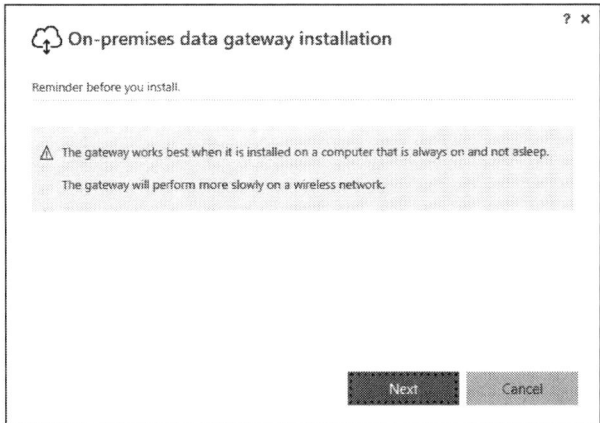

Figure 3.47: Reminder before installation

The **Getting ready to install the on-premises data gateway** page appears.

11. Specify the path of directory where you want to install the gateway in the **Install to** text box.

12. Click the **I accept the terms of use and privacy statement** checkbox.

13. Click the **Install** button, as shown in Figure 3.48.

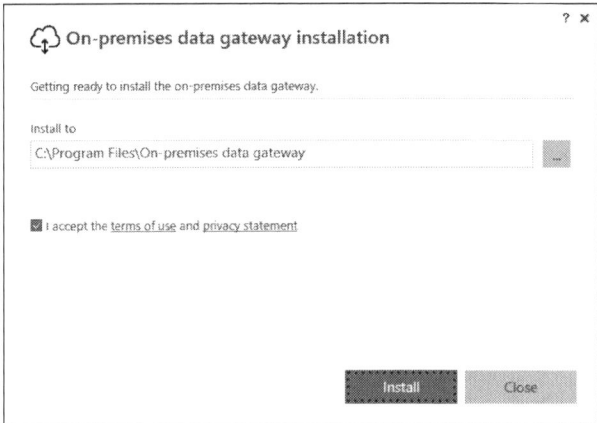

Figure 3.48: Specifying installation folder

The **Installing your on-premises data gateway** page appears providing status and progress of the installation, as shown in Figure 3.49.

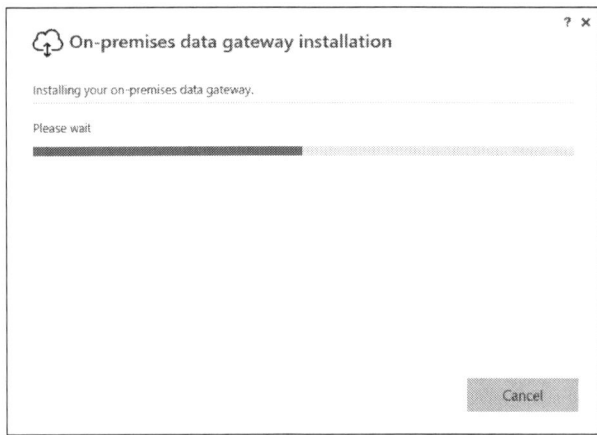

Figure 3.49: Progress of installation of the data gateway

The **Almost done** page of the **On-premises data gateway installation** wizard appears stating that installation was successful. This page also states that you need to sign in to register your gateway.

Configuring Gateway

Once the installation is successful, you need to configure the gateway so that it can be used with Power BI. For configuring the gateway, you need to continue with the steps followed in the earlier section "Downloading and Installing the On-Premises Data Gateway." Perform the following steps to configure/register gateway:

1. Enter the email address with which you want to register your gateway in the **Email address to use with this gateway** text box of the **On-premises data gateway** window.
2. Click the **Sign in** button, as shown in Figure 3.50.

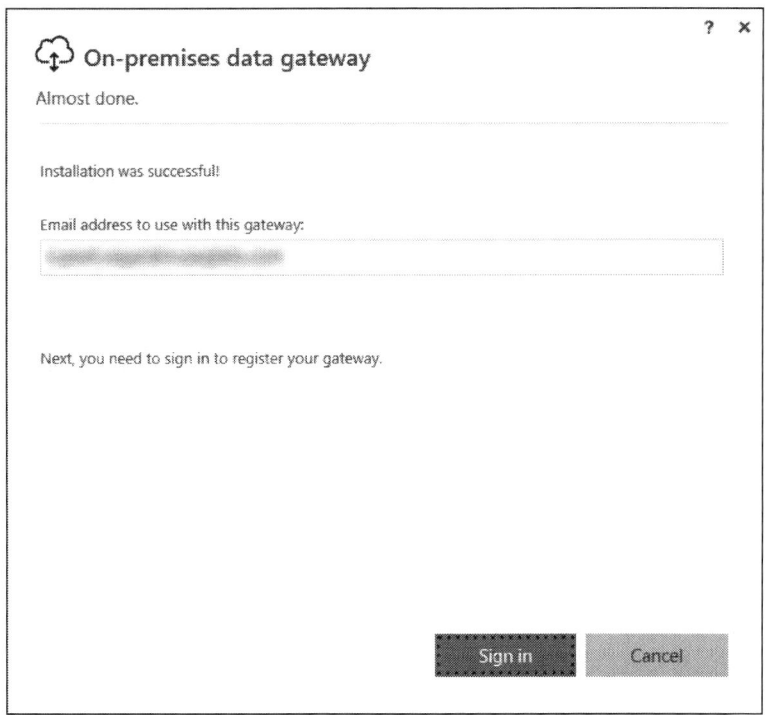

Figure 3.50: Specifying the email address

The **Sign in** window appears.

3. Enter the related password in the **Password** text box.
4. Click the **Sign in** button.

The next page of the **On-premises data gateway** window appears.

5. Select the **Register a new gateway on this computer** radio button to register a new gateway.
6. Click the **Next** button, as shown in Figure 3.51.

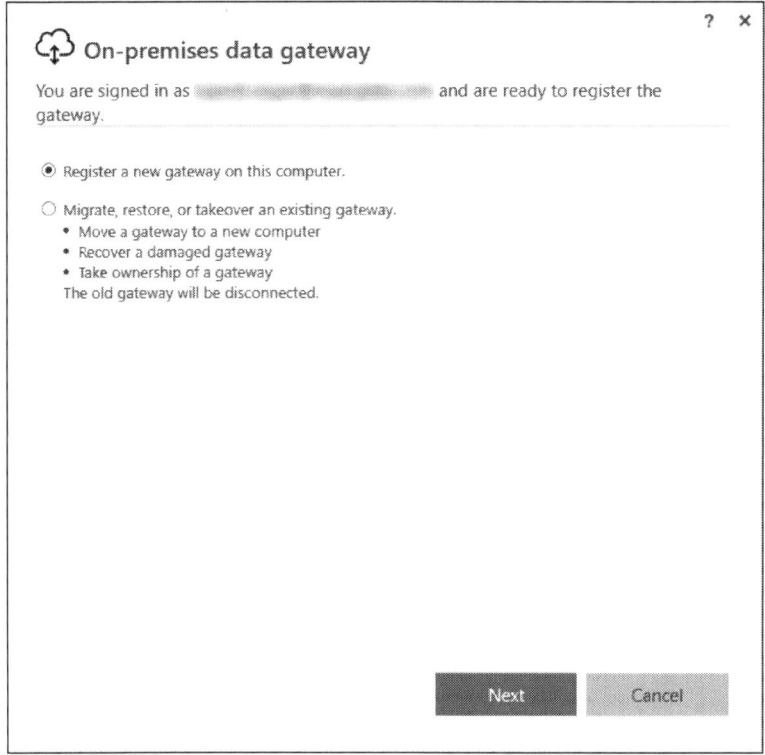

Figure 3.51: Specifying the type of gateway registration

The next page of the **On-premises data gateway** window asks you to specify name and recovery key for the gateway.

7. Enter the name of a gateway in the **New on-premises data gateway name** text box.
8. Enter the desired recovery key for the gateway in the **Recovery key** text box.
9. Enter the same recovery key in the **Confirm recovery key** text box.

10. Click the **Configure** button, as shown in Figure 3.52.

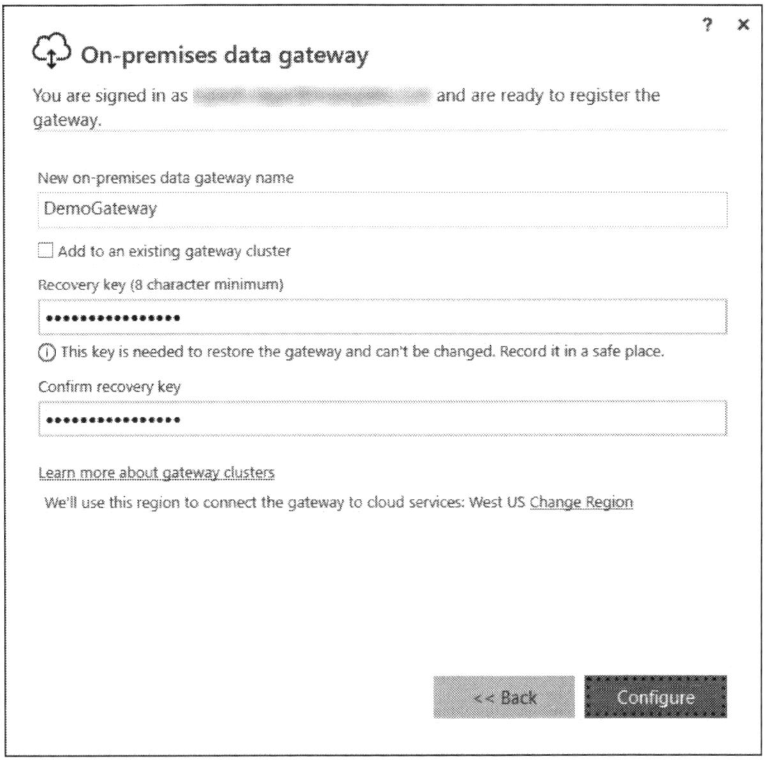

Figure 3.52: Configuring on-premises data gateway

Once the gateway is configured, you will see the status that **"The gateway Demo-Gateway is online and ready to be used."**

11. Click the **Close** button to close the **On-premises data gateway** window, as shown in Figure 3.53.

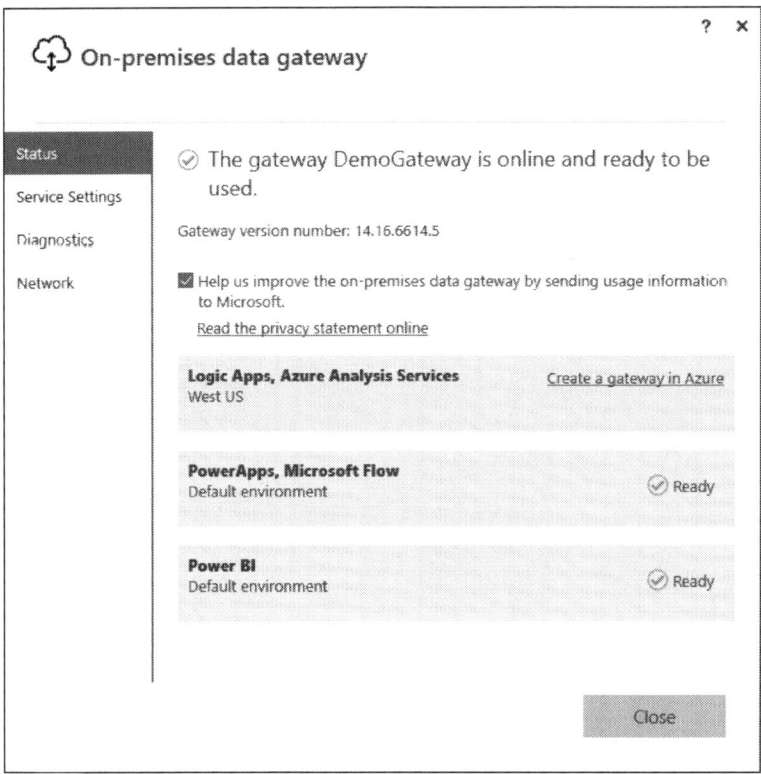

Figure 3.53: Completion status of gateway configuration

Adding a Data Source

After installing the on-premises data gateway, you can add a data source (SQL Server) to be used with the gateway. You can find the list of available gateways under the GATEWAY CLUSTERS section of the Gateways window. The settings related to the selected gateway cluster appear in the right pane of the Gateways window.

Perform the following steps to add a data source to the gateway created earlier:
1. Launch Power BI Service.
2. Click the **Settings** icon button. A list of options appears.
3. Click the **Manage gateways** option, as shown in Figure 3.54.

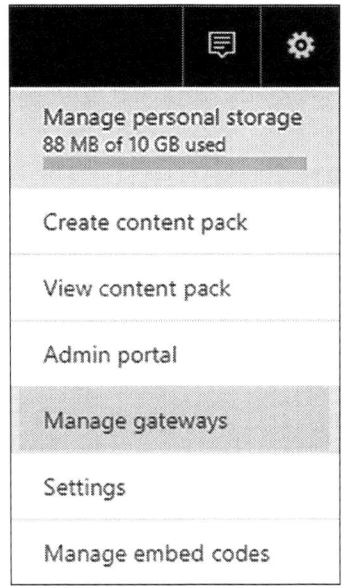

Figure 3.54: Clicking the Manage gateways option

The **Gateways** window appears. In the left pane, you will see a list of available gateways under the **GATEWAY CLUSTERS** section. The right pane displays information about the selected gateway and other settings related to the gateway under the **Gateway Cluster Settings** tab, as shown in Figure 3.55.

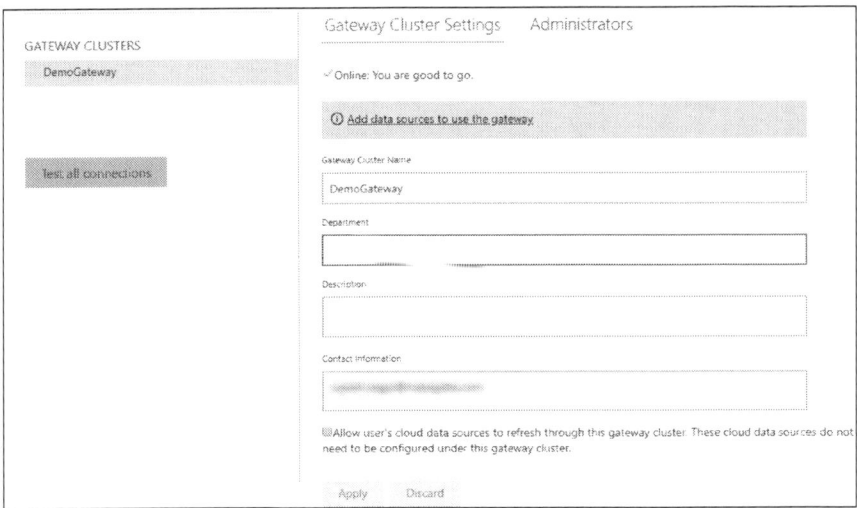

Figure 3.55: Gateway Cluster Settings

4. Click either **Add data sources to use this gateway** link or select the **Ellipsis** icon (...) next to the gateway and click the **ADD DATA SOURCE** option from the menu, as shown in Figure 3.56.

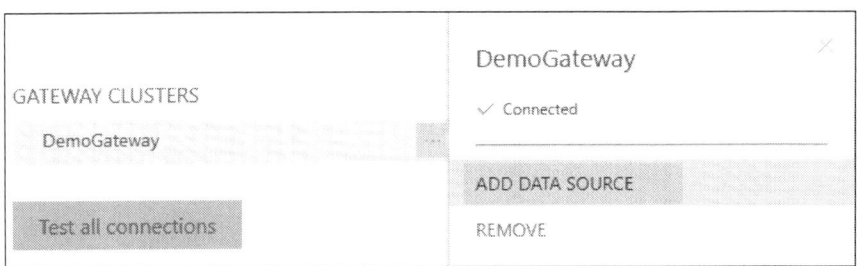

Figure 3.56: Clicking the ADD DATA SOURCE option

5. Enter the desired name for the data source in the **Data Source Name** text box.
6. Select the **SQL Server** option from the **Data Source Type** drop-down list to connect the gateway to the SQL Server data source.
7. Enter the name of SQL Server where the database is present in the **Server** text box.
8. Enter the same name of the database that you used earlier, in the **Database** text box.

9. Select the desired authentication method from the **Authentication Method** drop-down list, as shown in Figure 3.57.

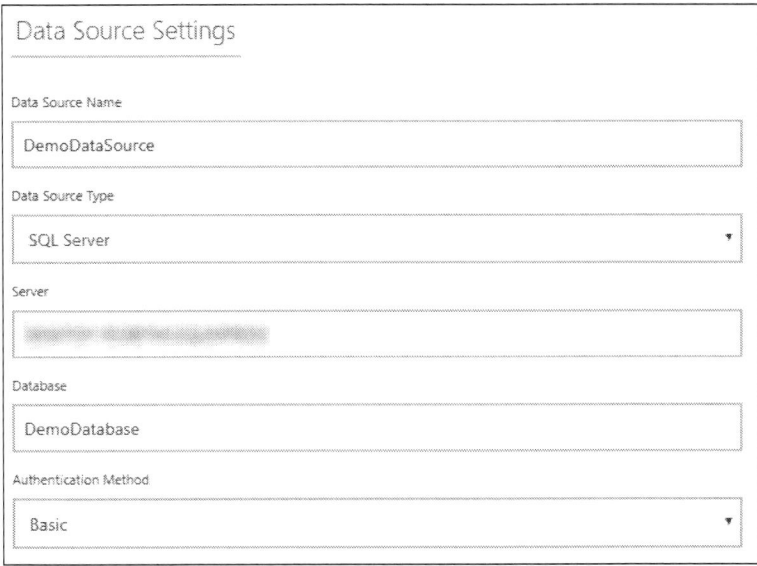

Figure 3.57: Specifying data source settings

After you select an authentication method, the **Username** and **Password** fields display.

10. Enter the relevant username in the **Username** text box.
11. Enter the applicable password in the **Password** text box.
12. Click the **Add** button, as shown in Figure 3.58.

Authentication Method

| Basic | ▾ |

The credentials are encrypted using the key stored on-premises on the gateway server. Learn more

Username

| 38 |

Password

| •••••••• |

> Advanced settings

| Add | Discard |

Figure 3.58: Specifying username and password

As you click the **Add** button, the **Connection Successful** message appears, which states that the connection has been established, as shown in Figure 3.59.

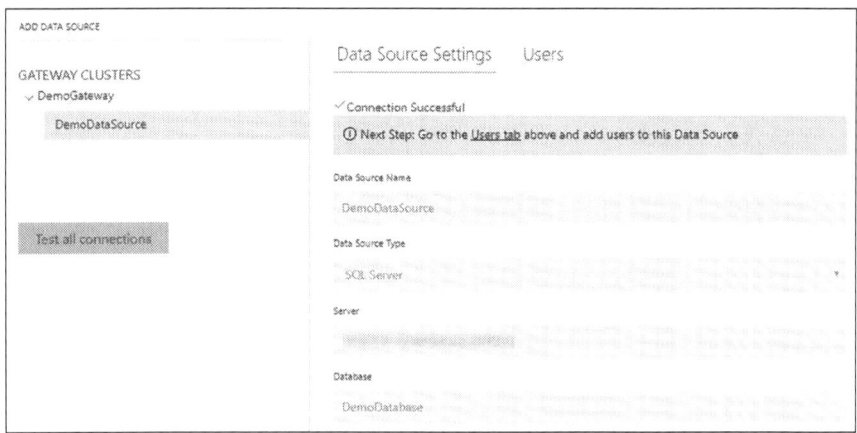

Figure 3.59: Connection Successful message

13. Select the **Users** tab to add users to the created data source.
14. Specify the email address of the person that you want to allow to publish reports using this data source.

15. Click the **Add** button. The specified person is added to the list box, as shown in Figure 3.60.

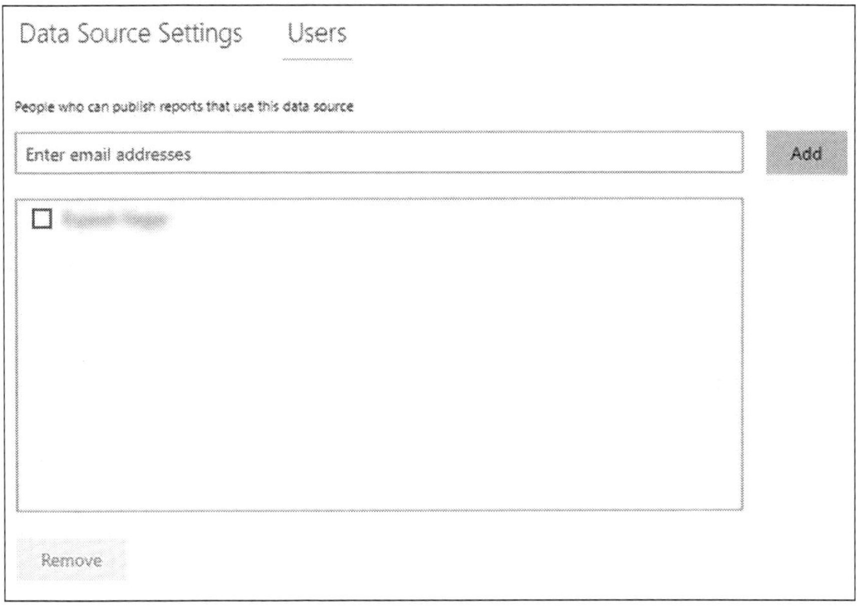

Figure 3.60: Adding users

Once the gateway is configured properly, the visuals start appearing on the report, as shown in Figure 3.61.

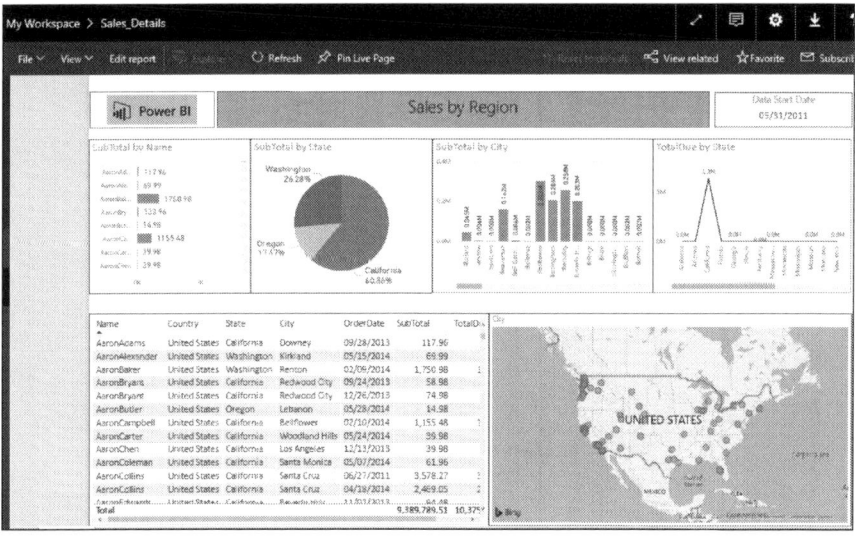

Figure 3.61: Displaying the report

Natural Language Querying

One of the prominent features of Power BI is that it supports natural language. You can ask questions in natural language and Power BI will answer these questions in the form of charts and graphs. This feature is commonly termed as Q&A.

Note

Q&A should not be replaced with a search engine because Q&A provides results from the data available in Power BI only while a search engine provides results from data available worldwide.

To use this feature, you should have some data in Power BI. If you do not have data, you can connect to the samples available in Power BI Service. Perform the following steps to connect to a sample:

1. Open Power BI Service.
2. Click the **Get Data** button in the left pane. The **Get Data** page appears in the right pane.
3. Click the **Samples** link, as shown in Figure 3.62.

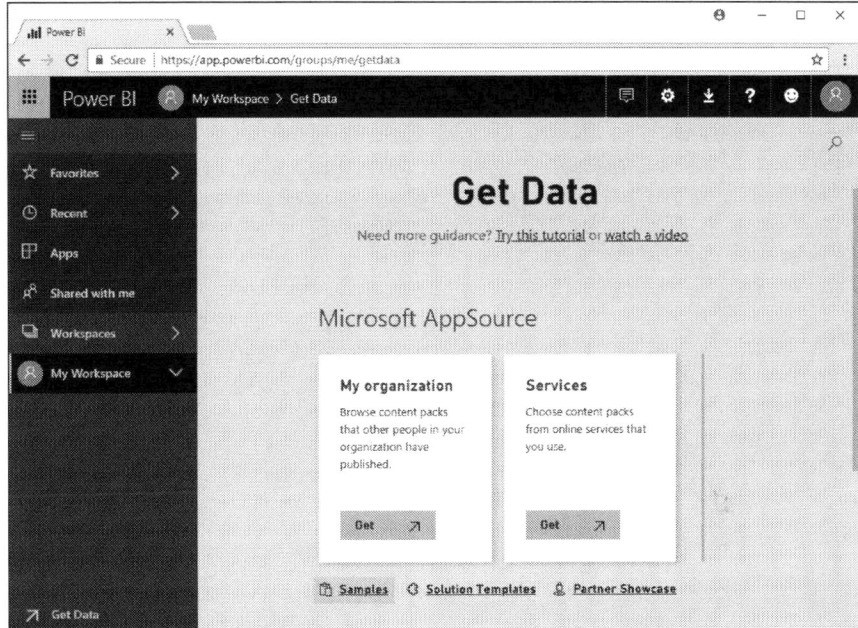

Figure 3.62: Clicking the Samples link

A list of available samples appears.

4. Select the desired sample that you want to use. In our case, we have selected **Sales and Marketing Sample**.

As you select a sample, a pane appears with the **Connect** button.

5. Click the **Connect** button, as shown in Figure 3.63.

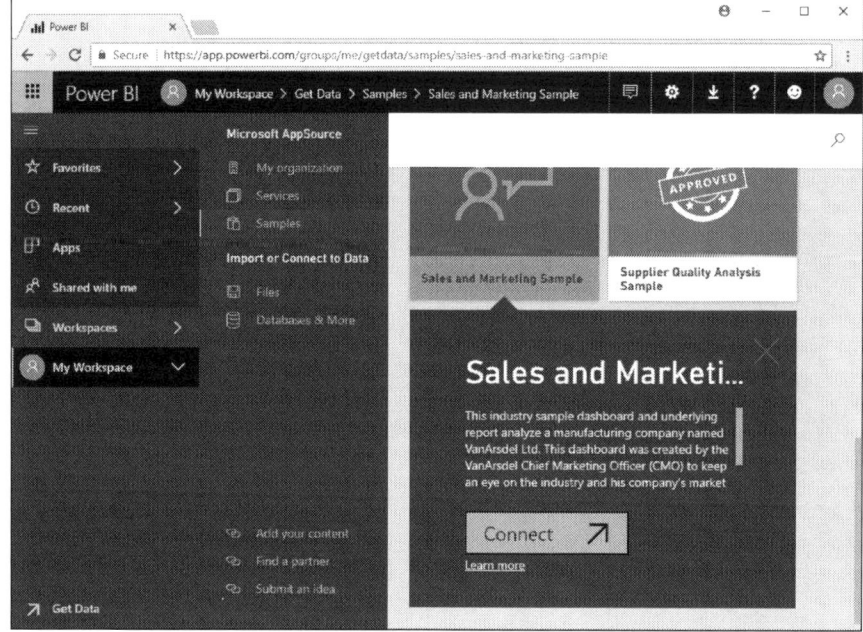

Figure 3.63: Connecting to a sample

The **Importing data** message box appears displaying the progress of importing data, as shown in Figure 3.64.

Figure 3.64: The Importing data message box

The **Success** dialog box appears containing the **Go to dashboard** button.

6. Click the **Go to dashboard** button, as shown in Figure 3.65.

Figure 3.65: The Success dialog box

The dashboard opens and displays the tiles pinned to the dashboard, as shown in Figure 3.66.

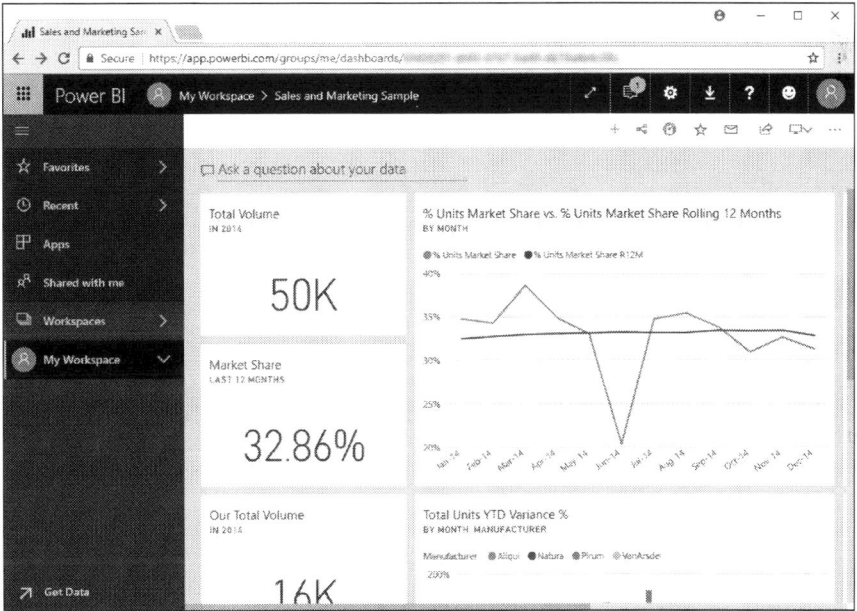

Figure 3.66: Displaying tiles on the dashboard

Perform the following steps to query Power BI in natural language:

1. Click on the **Ask a question about your data** text box. The **Q&A** page appears with a list of keywords that can be used for querying Power BI for a solution, as shown in Figure 3.67.

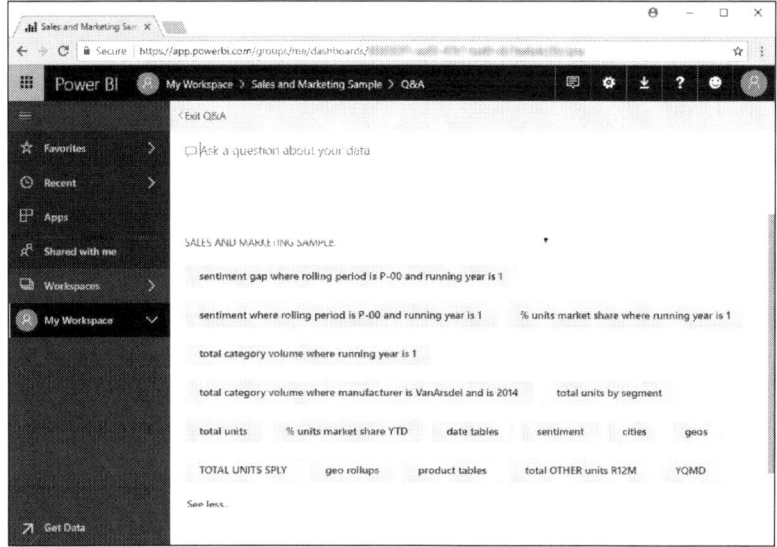

Figure 3.67: The Q&A page

2. Enter the desired query in the **Ask a question about your data** text box. In our case, we have entered **"count of product by manufacturer."** After we enter the keywords, we receive the result in the form of a visual, as shown in Figure 3.68.

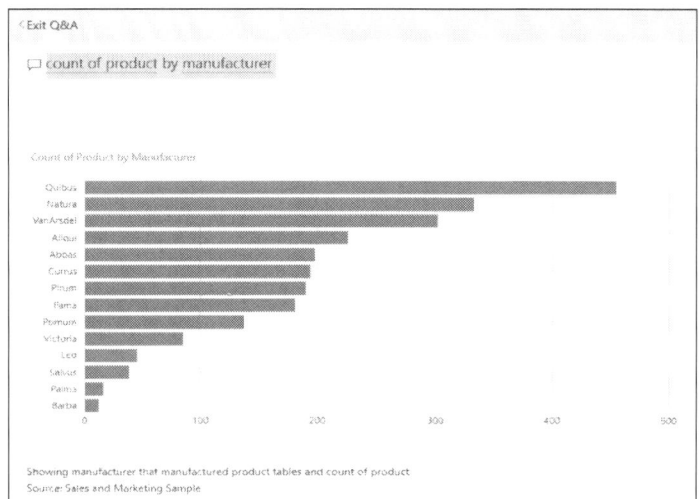

Figure 3.68: Viewing the result in the form of a visual

You can modify this visual as per your requirements. For example, to change the visual, you just need to select another visual from the VISUALIZATIONS pane and the data will be visualized in the updated visual, as shown in Figure 3.69.

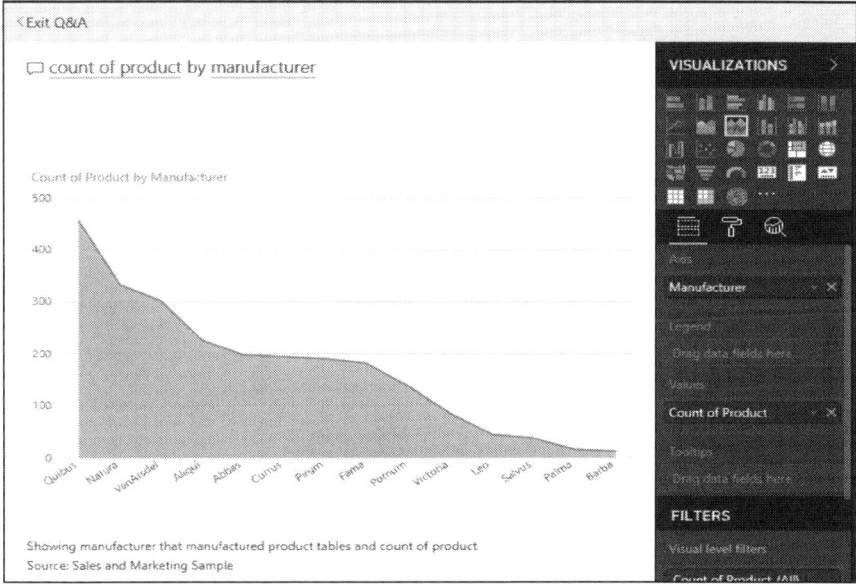

Figure 3.69: Viewing the updated visual

3. Click the **Pin visual** icon located in the upper right corner to pin this visual to the dashboard as a tile. The **Pin to dashboard** dialog box appears.
4. Select either the **Existing dashboard** or the **New dashboard** radio button to specify the dashboard where you want to pin this visual. In our case, we have selected **Existing dashboard**.
5. Select the desired dashboard from the drop-down list.
6. Click the **Pin** button, as shown in Figure 3.70.

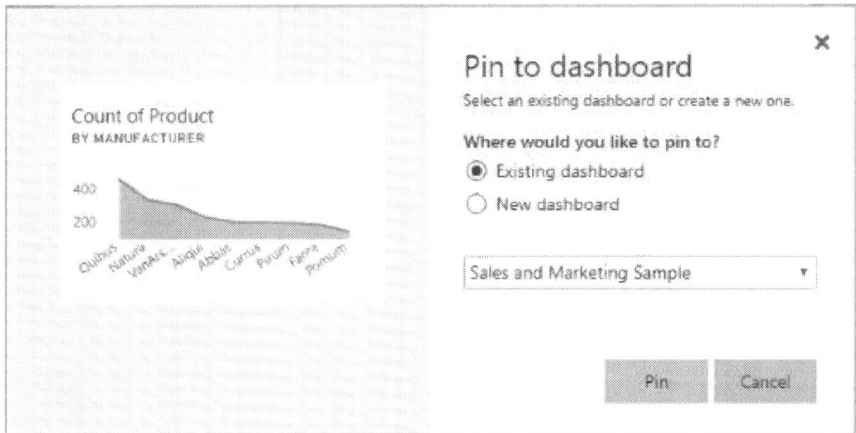

Figure 3.70: Pinning a visual to the dashboard

The **Pinned to dashboard** message box appears, which specifies that the created visual is pinned to the selected dashboard. It also contains the **Go to Dashboard** button that navigates you to the specified dashboard. The dashboard is now updated with the created visual, as shown in Figure 3.71.

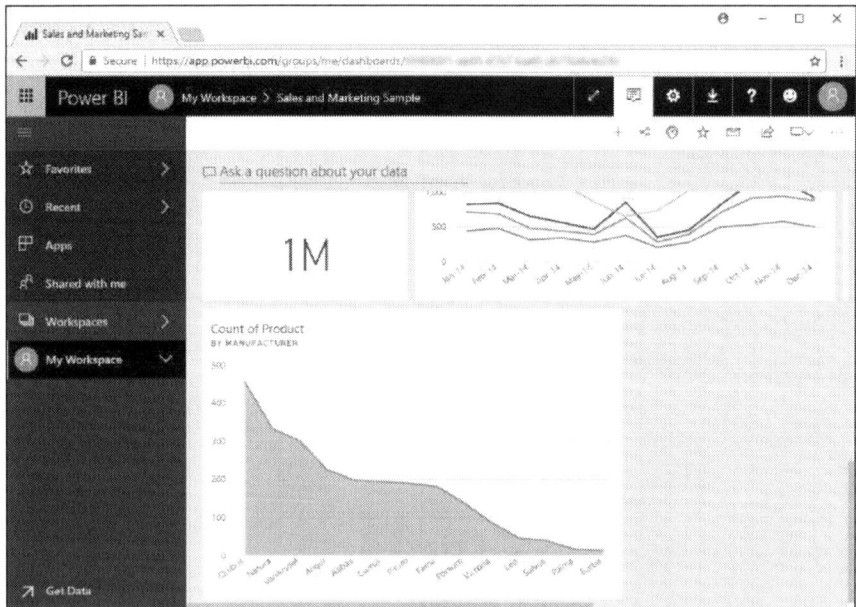

Figure 3.71: Displaying the dashboard

Data Refresh in Power BI

The point of using visuals to help communicate information about data being captured is to help users in making better business decisions. Because of this, you should always ensure that the data you are using for these Power BI reports is accurate and updated. However, it is not possible to import the entire data repeatedly into Power BI and create reports on that data. After creating Power BI reports with the data you have, you can set a scheduled refresh of that data so that your Power BI reports and visualizations reflect the updated information. This also eliminates the need to import the data repeatedly. You can also apply a manual refresh to the data by clicking the Refresh button in Power BI to visualize the updated data in your reports and dashboards.

In Power BI Service, a dataset is created automatically when:
1. Data is imported using the Get Data button from a content pack or file
2. The file is published from Power BI Desktop

The created dataset appears in My Workspace window in Power BI Service. You can perform different operations related to a dataset including exploring the data used in a report and setting up refresh by clicking the respective option from the menu that appears when you click the Ellipsis icon (...) next to a dataset name appearing under the Datasets tab.

Note
There could be multiple data sources for a single dataset.

A dataset contains:
- Information about the data sources
- Data source credentials
- A subset of data taken from the data source

Note
When you update data in the data source and want to apply the same update to your visuals in the report, you need to refresh data in Power BI. The data refresh updates the data in the dataset stored in Power BI. This type of data refresh is called a full refresh.

The data refresh process contains the following steps:
1. You use either **Refresh Now** or set up a refresh schedule for refreshing data in a dataset.

2. Power BI connects with the associated data sources by using the available information in the dataset.
3. Power BI queries the respective data sources for updated data.
4. Power BI loads the updated data into the dataset.
5. The visuals automatically update and display in your reports or dashboards based on the latest data.

Additional Information

Other than data refresh, there are a few additional types of refresh available in Power BI as follows:

1. **Package refresh:** Refers to a type of refresh that synchronizes your Power BI Desktop and Power BI Service and OneDrive. When you refresh data, the data remains at the original data source, only the dataset is updated as per the update made to OneDrive or SharePoint Online.
2. **Tile refresh:** Updates the visuals for tiles pinned on the dashboard whenever there is a change in data. Power BI checks for updated data every fifteen minutes. However, you can select the Refresh dashboard tiles option from the menu that appears when you click the Ellipsis icon (...) in a dashboard to apply tile refresh forcefully.
3. **Visual container refresh:** Updates the report visuals when you refresh the visual container.

Configuring Scheduled Refresh

You can apply a scheduled refresh by clicking the Schedule refresh icon or the Settings option from the Ellipsis icon(...) next to the dataset name. For successful configuration of a scheduled refresh, you need to set the following settings:

- Gateway connection
- Data source credentials
- Schedule refresh

Perform the following steps to configure a scheduled refresh for a dataset:

1. Launch Power BI Service.
2. Select the **My Workspace** option from the left pane.
3. Select the **Datasets** tab in the right pane. You will see a list of available datasets.
4. Click the **Schedule refresh** icon beside the desired dataset, as shown in Figure 3.72.

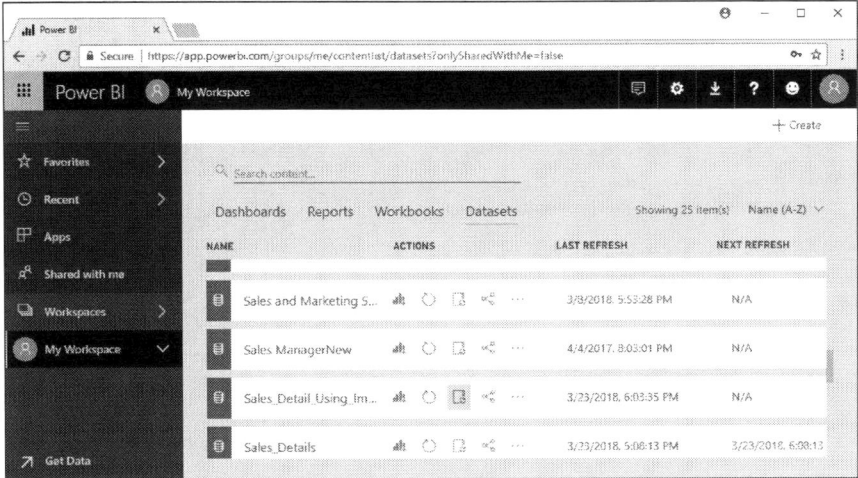

Figure 3.72: Clicking the Schedule refresh icon

The **Settings** page for the selected dataset appears.

5. Expand the **Gateway connection** node to display the settings related to a gateway connection. The related settings appear.

6. Select the **Use an on-prem data gateway** radio button. You will see the status as online.

7. Click the **Apply** button to apply the changes, as shown in Figure 3.73.

Settings for Sales_Detail_Using_Import

Refresh history

▲ Gateway connection

To use a data gateway, make sure the computer is online and the data source is added in Manage Gateways.

○ Use your data gateway (personal mode)

◉ Use an on-prem data gateway

Status	Department	Gateway	Contact information	Description
online		DemoGate...		

[Apply] [Discard]

▶ Data source credentials

▶ Parameters

▶ Scheduled refresh

Figure 3.73: Setting Gateway connection

8. Expand the **Data source credentials** node. Here, you will see a message stating that credentials are not required for Admin, as shown in Figure 3.74.

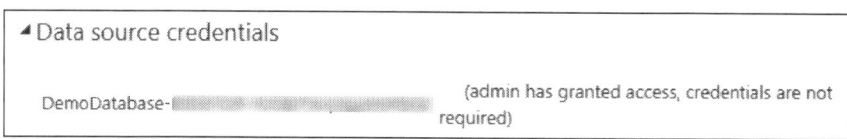

Figure 3.74: Expanding the Data source credentials node

9. Expand the **Scheduled refresh** node. This is where you need to specify settings related to a schedule refresh, including the frequency and time slots to refresh the dataset.

10. Drag the slider under the **Keep your data up to date** section to change its status to **On**.

11. Select the desired frequency for a refresh from the **Refresh frequency** drop-down list.

12. Select the desired time zone from the **Time zone** drop-down list.

13. Click the **Send refresh failure notification email to me** checkbox so that you receive an email if the refresh has failed.

14. Click the **Apply** button, as shown in Figure 3.75.

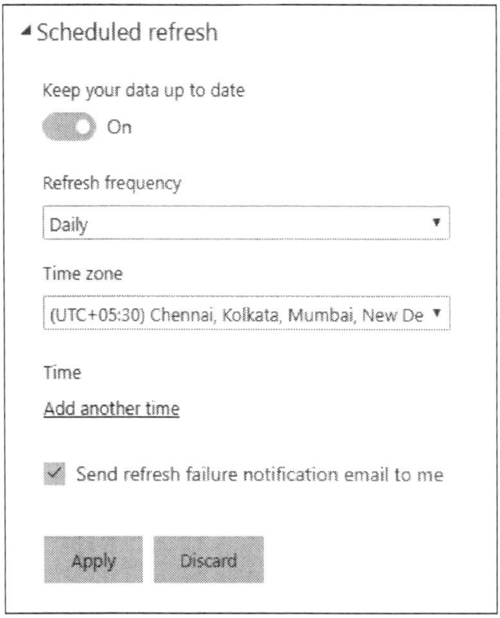

◢ Scheduled refresh

Keep your data up to date

⬤ On

Refresh frequency

| Daily ▼ |

Time zone

| (UTC+05:30) Chennai, Kolkata, Mumbai, New De ▼ |

Time
Add another time

☑ Send refresh failure notification email to me

Apply   Discard

Figure 3.75: Configuring scheduled refresh

After you click the Apply button, the scheduled refresh is configured.

When the data is loaded through DirectQuery, a direct connection is established between Power BI and the database. Power BI thus queries the database directly whenever you interact with any visual on a report. The refresh frequency for the dataset created using DirectQuery is set to 1 hour by default. However, you can change it as per your requirements by changing the frequency in the Refresh frequency drop-down list and then clicking the Apply button, as shown in Figure 3.76.

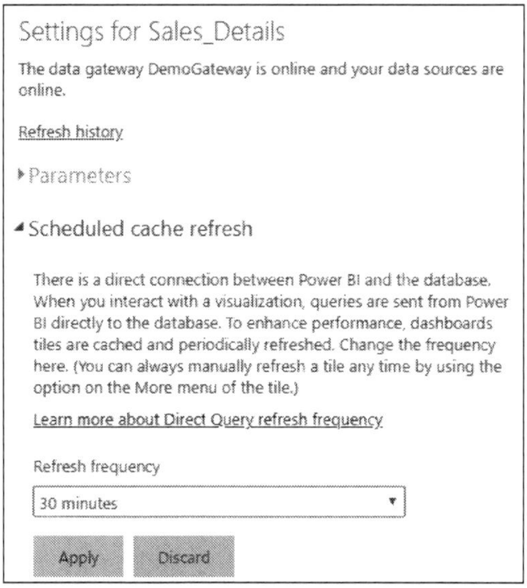

Figure 3.76: Setting refresh frequency

You can also see the refresh history by clicking the Refresh history link. This opens the Refresh history dialog box and displays the history of data refresh, as shown in Figure 3.77.

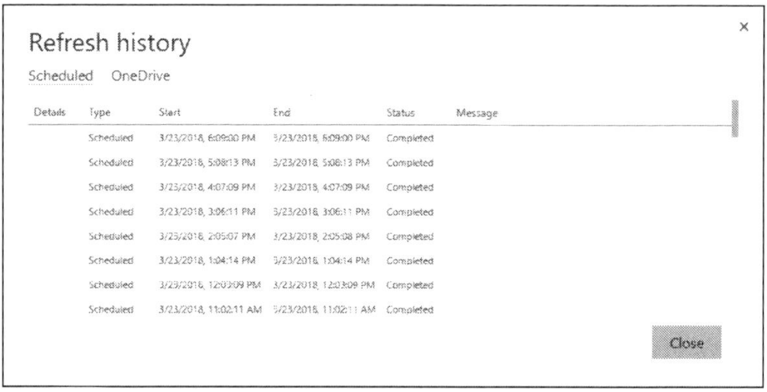

Figure 3.77: The Refresh history dialog box

Creating Content Packs

A content pack is a complete package of your dashboard, report, and dataset that can be shared with other users in your organization. You can create the content pack and publish it to the team. After you publish the content pack, it becomes available in a centralized repository called **AppSource**. This repository helps team members easily locate reports and datasets published for them.

Note

The reports and datasets published as content packs possess Power BI features, including support for data exploration, data refresh, visuals, and Q&A.

You can locate content packs in the central repository only when you are a member of a group, such as the entire organization, distribution group, security group, or Office 365 group, to which the content pack is published. You should note that the content pack data is read-only for all members of the group. However, you can copy reports to use them as a base for a personalized version of the content pack.

Note

You should have a Power BI Pro account for creating and accessing an organizational content pack.

Perform the following steps to create and publish a content pack:
1. Launch Power BI Service.
2. Click the **Settings** icon button. A drop-down menu appears.
3. Click the **Create content pack** button in the drop-down menu, as shown in Figure 3.78.

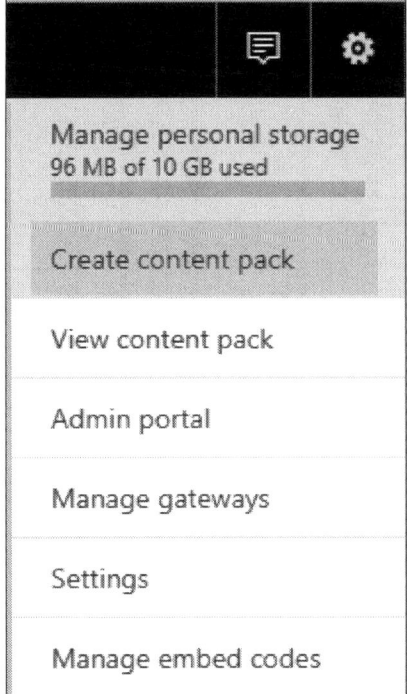

Figure 3.78: Creating a content pack

The **Create content pack** page appears.

4. Select either the **Specific groups** radio button to allow a specific group to access this content pack or select the **My entire organization** radio button to allow the entire organization to access this content pack.
5. Enter the desired title for the content pack in the **Title** text box.
6. Enter the desired description for the content pack in the **Description** text box.
7. Click the **Upload** icon to upload an image for the content pack, as shown in Figure 3.79.

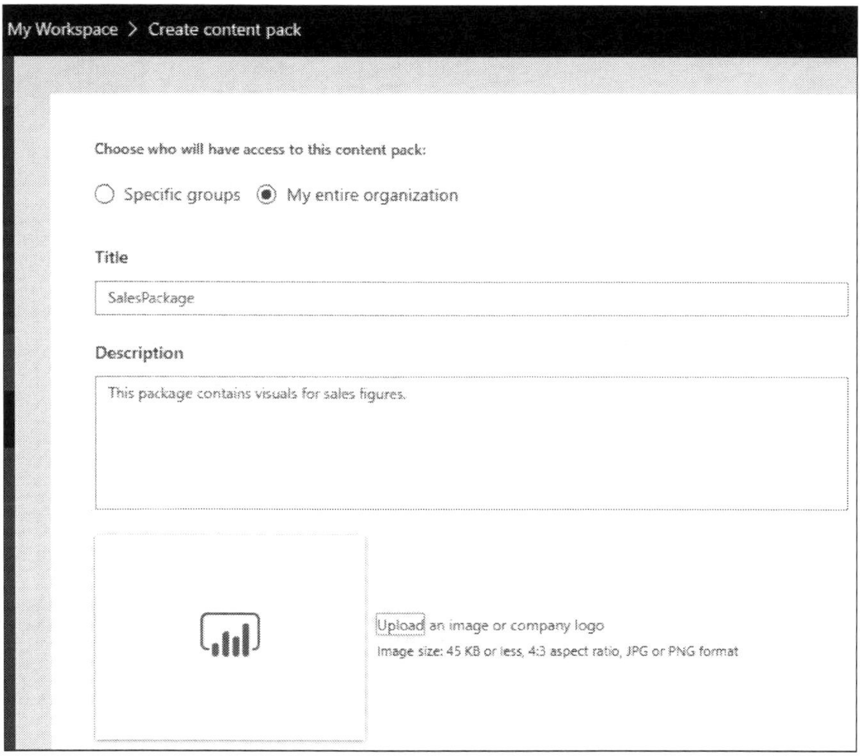

My Workspace > Create content pack

Choose who will have access to this content pack:

○ Specific groups ◉ My entire organization

Title

SalesPackage

Description

This package contains visuals for sales figures.

Upload an image or company logo
Image size: 45 KB or less, 4:3 aspect ratio, JPG or PNG format

Figure 3.79: Clicking the Upload link

The **Open** dialog box appears.

8. Navigate to the location where the image is located.
9. Select the image.
10. Click the **Open** button to upload the image, as shown in Figure 3.80.

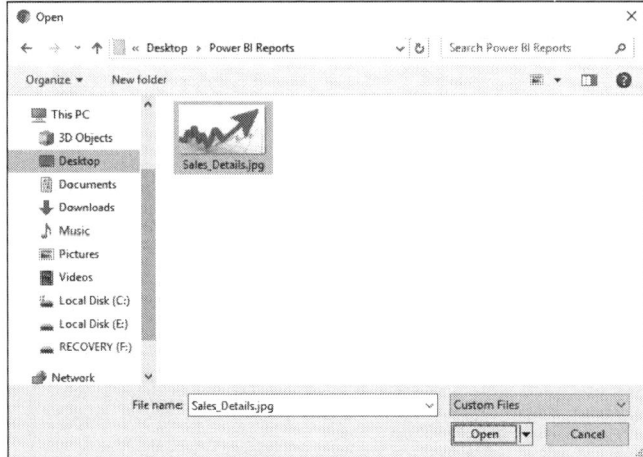

Figure 3.80: Uploading an image

The selected image is uploaded.

11. Select the desired dashboard to be published in the **Dashboards** list box. The related report and dataset are automatically selected from the **Reports** and the **Datasets** list boxes, respectively.

12. Click the **Publish** button, as shown in Figure 3.81.

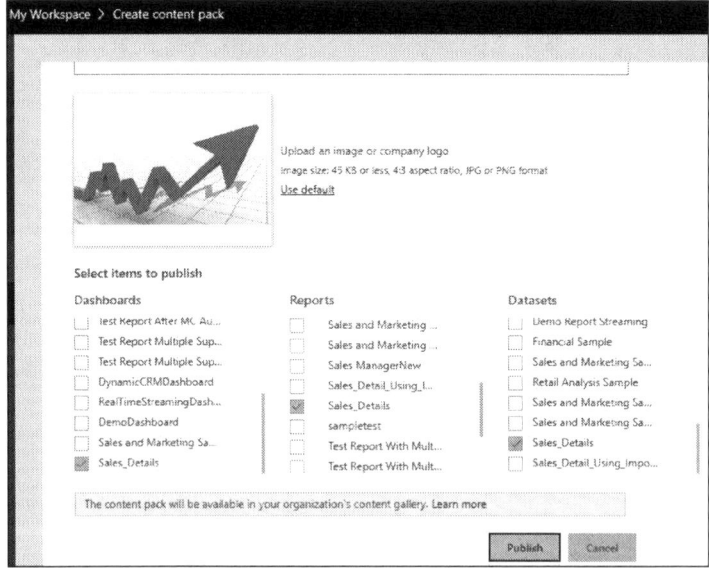

Figure 3.81: Publishing a content pack

A pop-up appears stating that the content pack is published successfully and added to the organization's content gallery.

Power BI Integration with Cortana Suite

As you know, Cortana is a feature of Windows 10 that gives relevant results to your queries made in natural language. Cortana can be integrated with Power BI to provide relevant information directly from Power BI dashboards and reports. When you integrate Cortana with Power BI, Cortana also looks into Power BI dashboards and reports for the related keywords each time you query Cortana.

To integrate Power BI with the Cortana suite, you need the following:
- A system running Windows 10 with version 1511 or later
- The Cortana feature to be turned on
- A Power BI account
- An Azure Active Directory (Azure AD)/Work or School account
- One or more datasets configured such that they can be used with Cortana

Creating a Cortana Answer Page and Publishing It

While integrating Power BI with Cortana, it is advisable to set the size of the report specifically for Cortana, which is called the Cortana answer page.

Perform the following steps to create a Cortana answer page:
1. Open the desired PBIX file that you want to use as a Cortana answer page.
2. Click the **Format** icon.
3. Expand the **Page Size** category.
4. Select the **Cortana** option from the **Type** drop-down list under the **Page Size** category.
5. Expand the **Page Information** category.
6. Enter the desired name for the report in the **Name** text box.
7. Drag the slider to change the status of the **Q&A** option to **On**.
8. Enter the keywords/alternate names in the text box below the **Q&A** option, as shown in Figure 3.82.

Figure 3.82: Creating a Cortana answer page

9. Click the **Save** button to save your changes.
10. Click the **Publish** button under the **Share** section to publish the created report.

The **Sign in** dialog box displays.

11. Enter the email address in the **Sign in** text box.
12. Click the **Sign in** button, as shown in Figure 3.83.

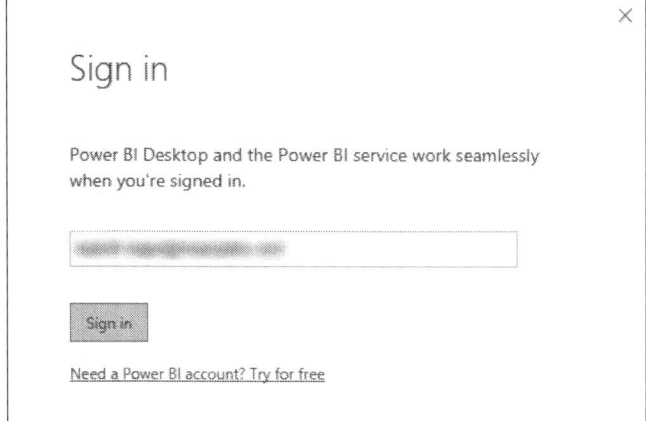

Figure 3.83: The Sign in dialog box

The **Sign in to your account** dialog box appears.

13. Enter the relevant password in the **Enter password** text box.
14. Click the **Sign in** button, as shown in Figure 3.84.

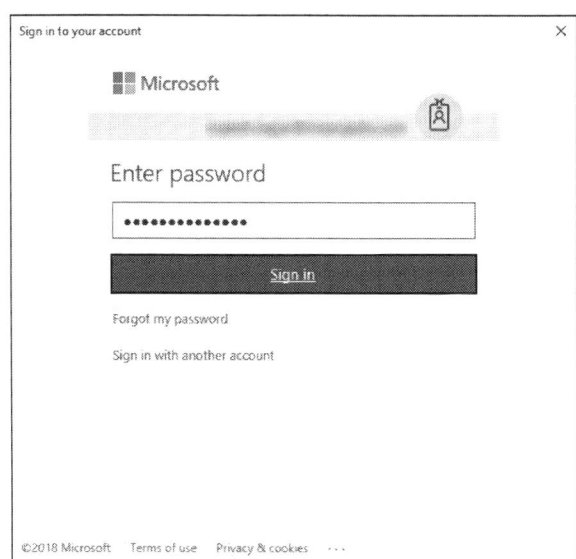

Figure 3.84: The Sign in to your account dialog box

The **Publish to Power BI** dialog box appears.

15. Select the desired destination from the **Select a destination** list box.

16. Click the **Select** button, as shown in Figure 3.85.

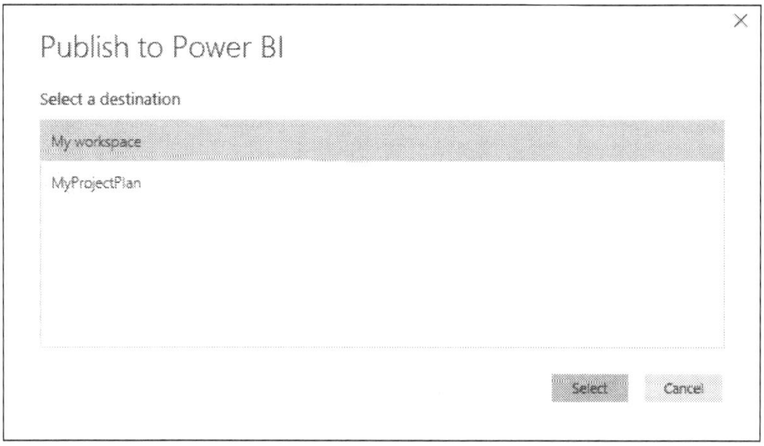

Figure 3.85: The Publish to Power BI dialog box

The **Publishing to Power BI** dialog box appears.

17. Click the **Got it** button to close the dialog box.

The report has now been published in Power BI Service.

Enabling Cortana to Access the Dataset

Cortana can easily access Power BI reports. For this, you need to select the "Enable Cortana to access this dataset" checkbox to enable the related dataset of the report. It allows users with access to the dataset to get answers from Cortana.

Perform the following steps to enable Cortana to access the dataset:

1. Browse the following link:
   https://app.powerbi.com/

The **Sign in to your account** page appears.

2. Enter the email address that you used with Power BI Desktop for publishing the report in the **Sign in** text box.
3. Click the **Next** button, as shown in Figure 3.86.

Figure 3.86: The Sign in to your account page

4. Enter the relevant password in the **Enter password** text box.
5. Click the **Sign in** button, as shown in Figure 3.87.

Figure 3.87: Signing in to Power BI Service

6. Select the **My Workspace** option from the left pane.

7. Select the **Reports** tab in the right pane. The available reports are displayed under the **Reports** tab.
8. Click the **View related** ( ) icon next to the name of the report that you want to allow Cortana to access, as shown in Figure 3.88.

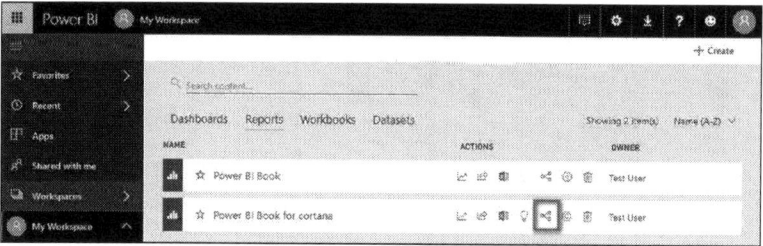

Figure 3.88: Clicking the View related icon

The **Related content** pane appears displaying the dashboards and datasets, as shown in Figure 3.89.

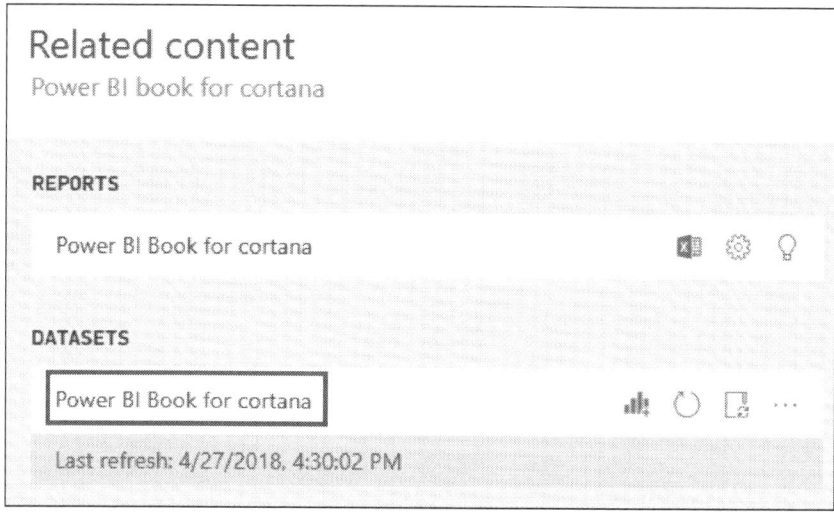

Figure 3.89: The Related content pane

9. Select the **Ellipsis** icon (...). A drop-down menu appears.
10. Select the **Settings** option from the drop-down menu, as shown in Figure 3.90.

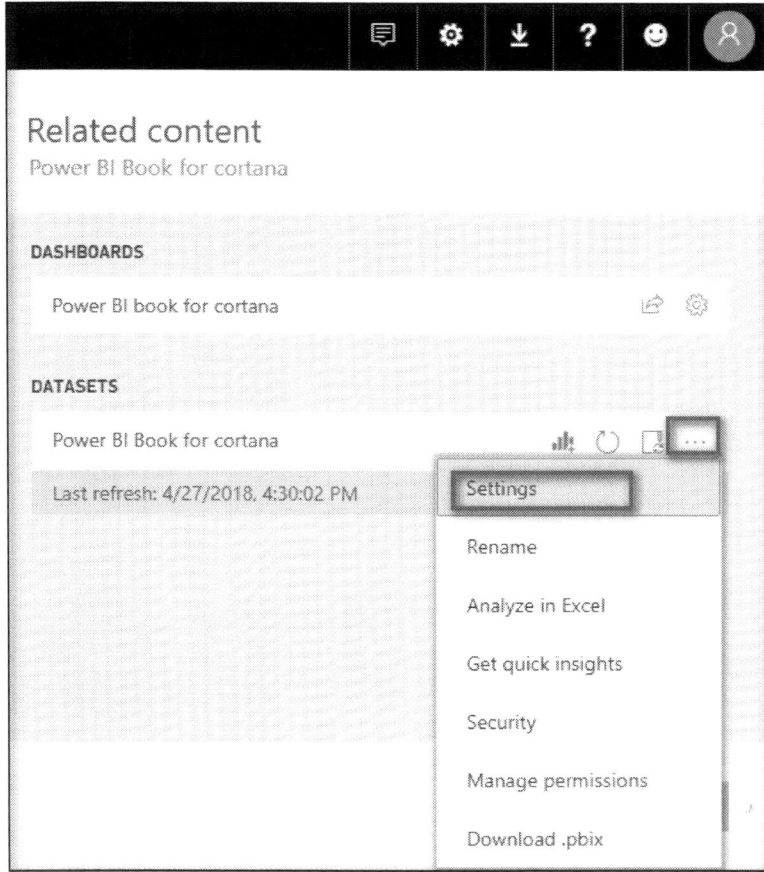

Figure 3.90: Selecting the Settings option

The **Settings** page appears with the related settings of the selected dataset.

11. Expand the **Q&A and Cortana** option.
12. Select the **Allow Cortana to access this dataset** checkbox.
13. Click the **Apply** button, as shown in Figure 3.91.

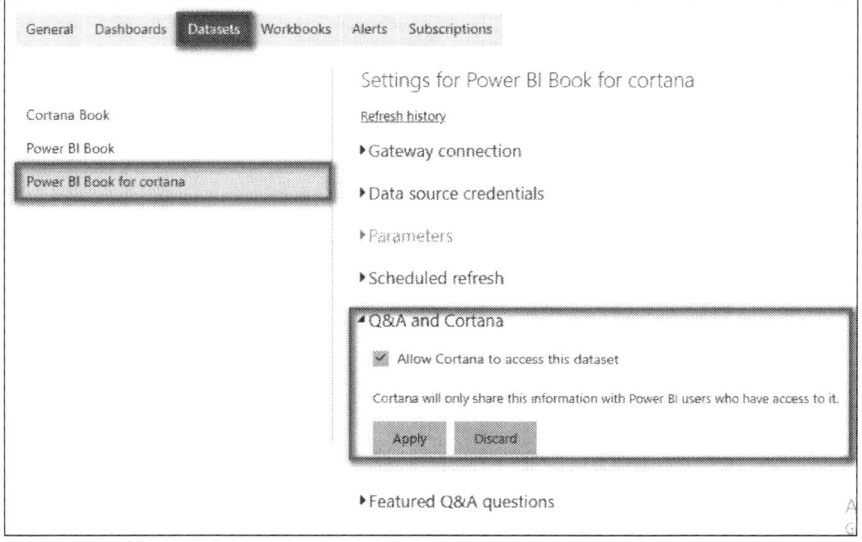

Figure 3.91: Allowing Cortana to access the dataset

Cortana is now allowed to access the dataset.

Adding Power BI Credentials to Windows 10

As discussed earlier, Cortana is a feature of Windows 10. Therefore, you need to connect to Windows 10 through Power BI credentials for integrating Power BI with Cortana. You just need Windows 10 with version 1511 or later.

Note

You can check the version of Windows 10 by navigating to Start→ Settings→ System→ About. The About page shows the Windows Edition and Version under the Windows specifications section.

Perform the following steps to add Power BI credentials to Windows 10:

1. Click the **Start** button. The **Start** menu appears.
2. Click the **Settings** icon, as shown in Figure 3.92.

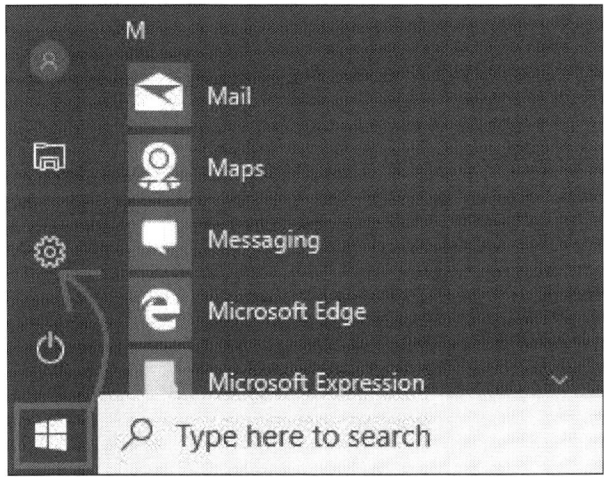

Figure 3.92: Clicking the Settings icon

The **Settings** window appears.

3.   Click the **Accounts** link, as shown in Figure 3.93.

Figure 3.93: Clicking the Accounts link

A page with related account settings appears.

4. Select the **Access work or school** tab under the **Accounts** section. The **Access work or school** page appears in the right pane.

5. Click the **Connect** button, as shown in Figure 3.94.

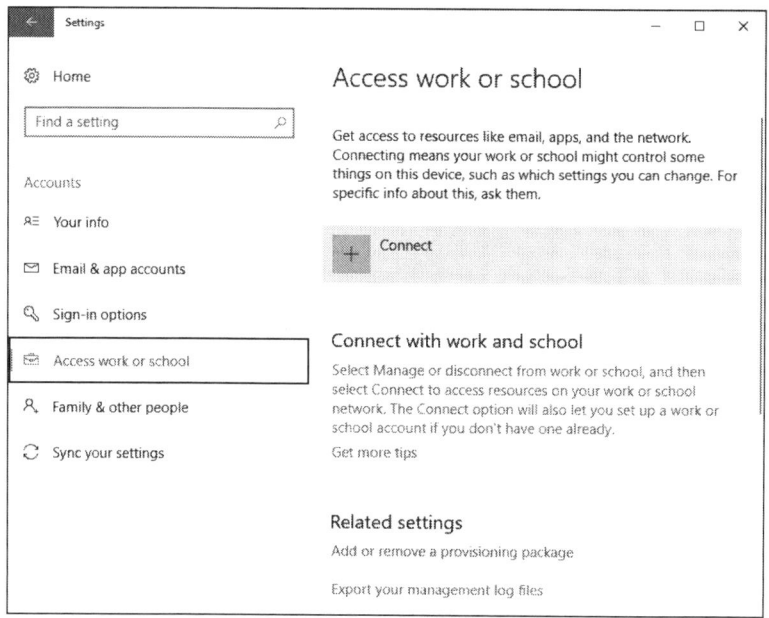

Figure 3.94: Clicking the Connect button

The **Microsoft account** window appears.

6. Click the **Join this device to Azure Active Directory** link, as shown in Figure 3.95.

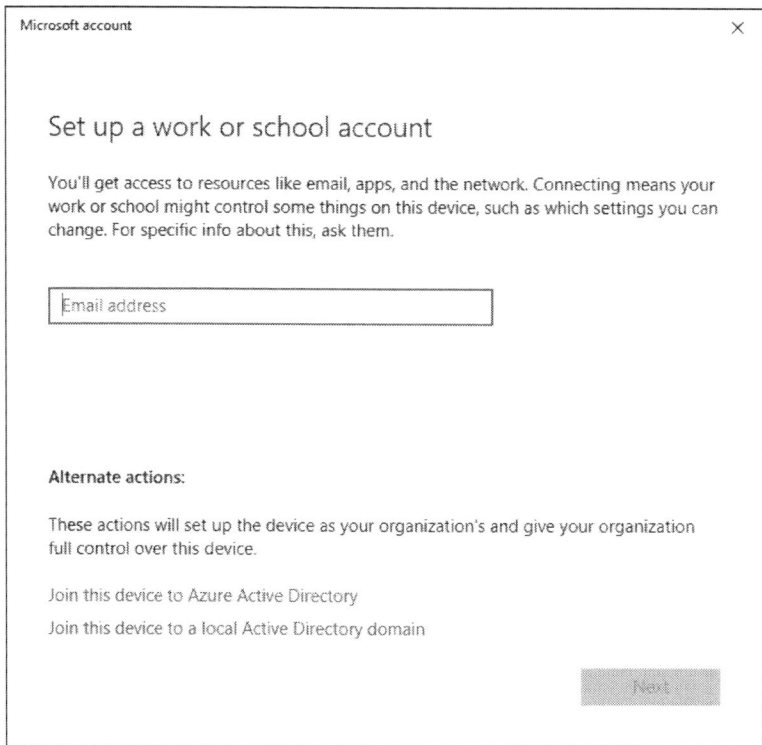

Figure 3.95: Clicking the Join this device to Azure Active Directory link

The **Sign in** page appears.

7.   Enter the email address in the **Email address** text box.
8.   Click the **Next** button.

The next page appears.

9.   Enter the applicable password in the **Enter password** text box.
10.   Click the **Sign in** button.

You are connected with the specified credentials, as shown in Figure 3.96.

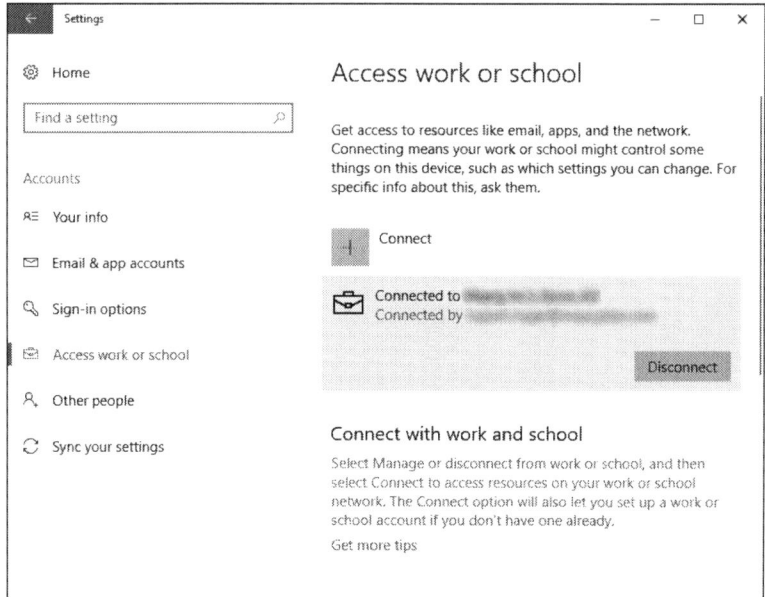

Figure 3.96: Displaying the new account

Accessing the Report through Cortana

Once you have created the new account, you can switch to Power BI account to access Power BI reports through Cortana, as shown in Figure 3.97.

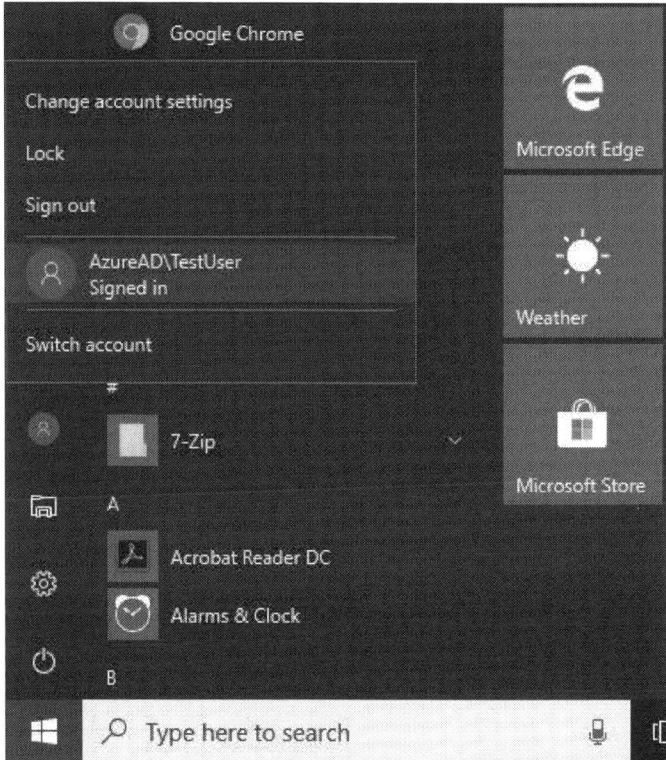

Figure 3.97: Switching to Power BI account

Perform the following steps to access the Power BI report through Cortana:

1. Press the **Windows** + **S** key combination on the keyboard.
2. Type the desired keywords in the **Type here to search** text box. In our case, we have typed **Leads revenue by Employee**.

As we type the keywords, the associated Power BI report is shown in the Cortana Search box, as shown in Figure 3.98.

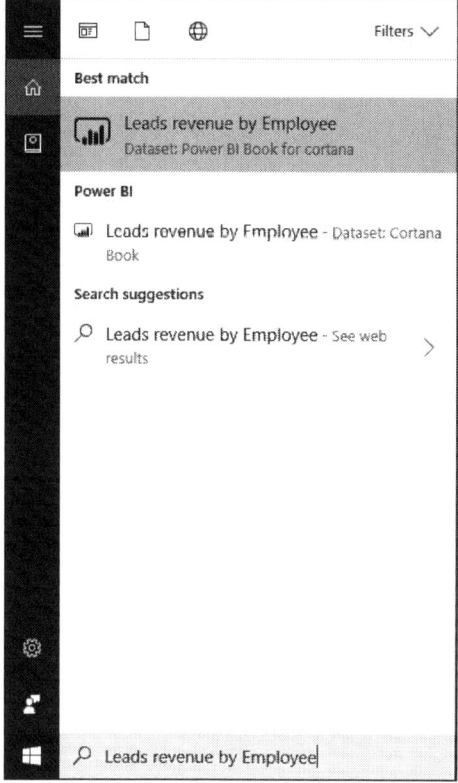

Figure 3.98: Displaying the search result

3. Click the report. The visuals of the report are shown in the Cortana Search box, as shown in Figure 3.99.

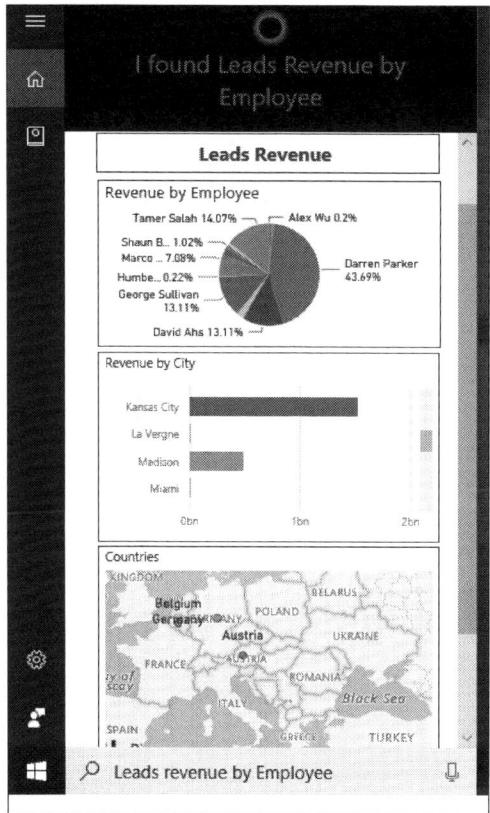

Figure 3.99: Displaying the visuals of the report

4. Click the **Open in Power BI** link to open the report into Power BI, as shown in Figure 3.100.

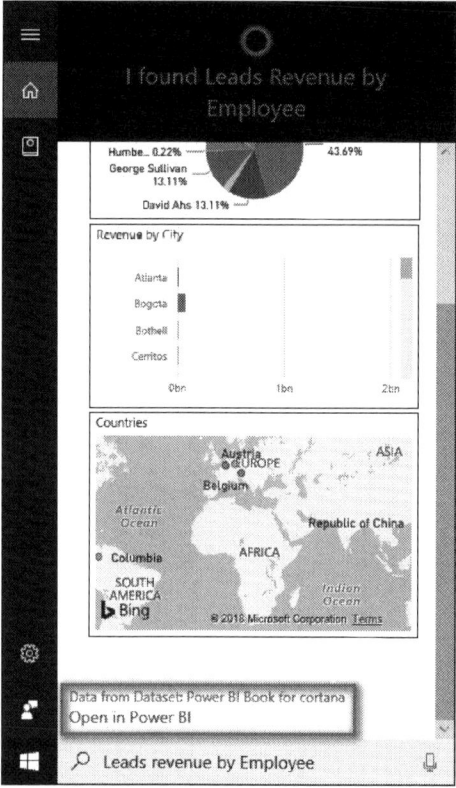

Figure 3.100: Clicking the Open in Power BI link

The selected report opens in Power BI Service, as shown in Figure 3.101.

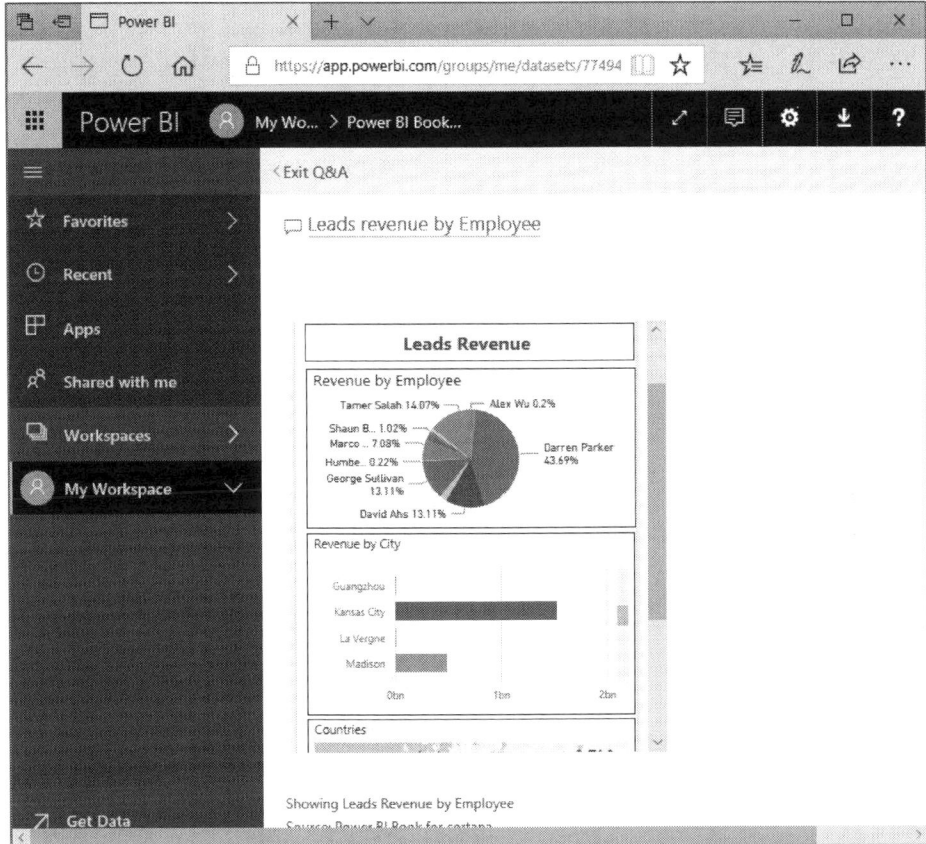

Figure 3.101: Opening the report in Power BI Service

Summary

Power BI is a business intelligence and reporting tool that allows users to create intuitive reports. It supports many data sources including Excel, SQL Server, PostgreSQL, Dynamics CRM, and MySQL. This chapter focused on integrating Power BI with SQL Server. The two options, that is, using the Import option and the DirectQuery option, for getting data into Power BI from SQL Server were discussed. Both the options of establishing a relationship between tables were also elaborated. This chapter provided in-depth information about using DAX expressions and creating calculated columns and tables. You also became familiar with the process of creating a report wherein data was loaded to Power BI through the Import option and the DirectQuery option. A gateway, a software that allows users to access data located on an on-premises system or network so that it can be used in a cloud service later, was set up on the machine on which SQL Server was installed. This chapter also provided in-depth information about data refresh in Power BI and provided a step-by-step procedure for creating a content pack. The last part of the chapter provided detailed information about the integration of Power BI with Cortana.

Chapter 4
Power BI on Open Source Stack

PostgreSQL is an open source object-relational database management system (ORDBMS). It is based on POSTGRES, which was introduced at Berkeley Computer Science Department at the University of California. The idea behind using the name PostgreSQL was to highlight the relationship between POSTGRES and SQL. This means that a large set of SQL standards are supported by PostgreSQL. In addition, PostgreSQL also supports:

- Triggers
- Views
- Complicated queries
- Foreign keys
- Transactional integrity
- Concurrency control

This chapter provides a step-by-step guide to integrating Power BI with PostgreSQL and creating reports in Power BI through the PostgreSQL database.

Integrating PostgreSQL with Power BI

As discussed in Chapter 1, Power BI is a business intelligence and reporting tool that can be integrated with hundreds of data sources. PostgreSQL is the most advanced open source database available today. You can integrate PostgreSQL with Power BI to analyze and visualize data and create interactive reports based on the available data.

To integrate PostgreSQL with Power BI, you will need the following:
- PostgreSQL
- PostgreSQL database
- Npgsql connector
- Power BI Desktop

Downloading and Installing Npgsql Connector

Npgsql is an open source ADO.NET Data Provider or connector that allows Power BI users to connect to PostgreSQL database.

DOI 10.1515/9781547400720-004

Perform the following steps to download and install the Npgsql connector:

1. Visit the following link to download the latest version of Npgsql:
   https://github.com/npgsql/Npgsql/releases
2. Click the **Npgsql-3.2.7.msi** link to download the .msi file.

The selected file will download and display in the **Downloads** folder on your computer.

3. Navigate to the **Downloads** folder and locate the downloaded file.
4. Double-click the **Npgsql-3.2.7.msi** file.

The **Welcome to the Npgsql 3.2.7 Setup Wizard** page of the **Npgsql 3.2.7 Setup** wizard appears.

5. Click the **Next** button to start the installation, as shown in Figure 4.1.

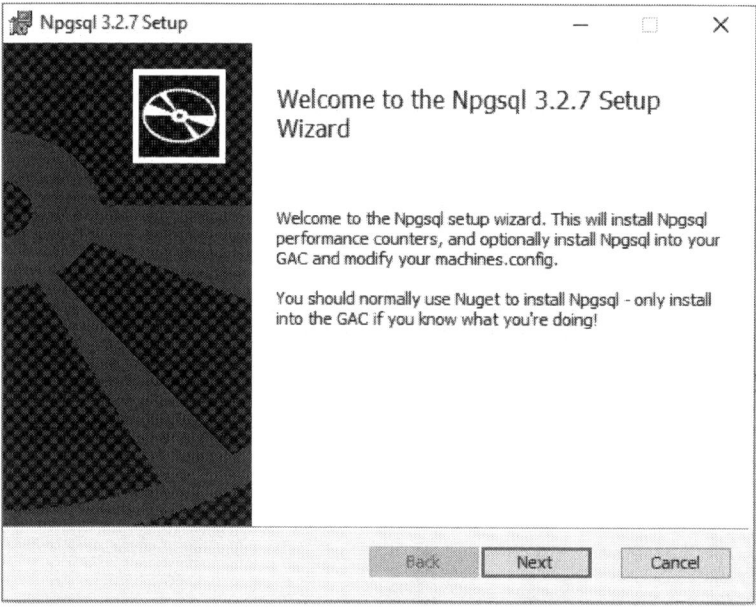

Figure 4.1: The Npgsql 3.2.7 Setup wizard

The **End-User License Agreement** page appears.

6. Click the **I accept the terms in the License Agreement** checkbox.
7. Click the **Next** button to continue the installation, as shown in Figure 4.2.

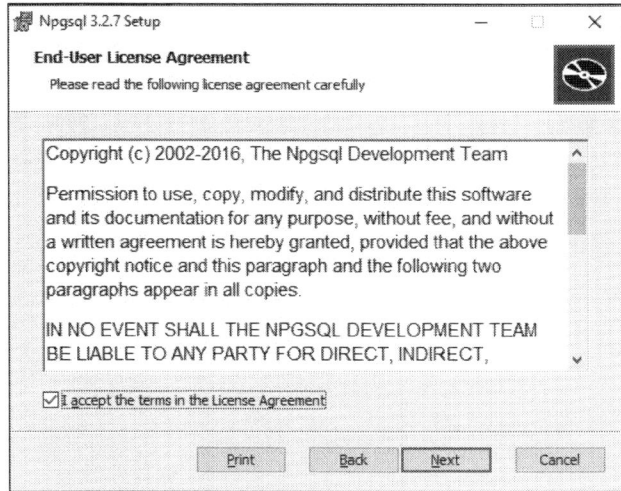

Figure 4.2: Accepting the license agreement

The **Custom Setup** page appears.

8. Click the **Npgsql GAC Installation** button. A list of options appears.
9. Select the **Entire feature will be installed on local hard drive** option.
10. Click the **Next** button, as shown in Figure 4.3.

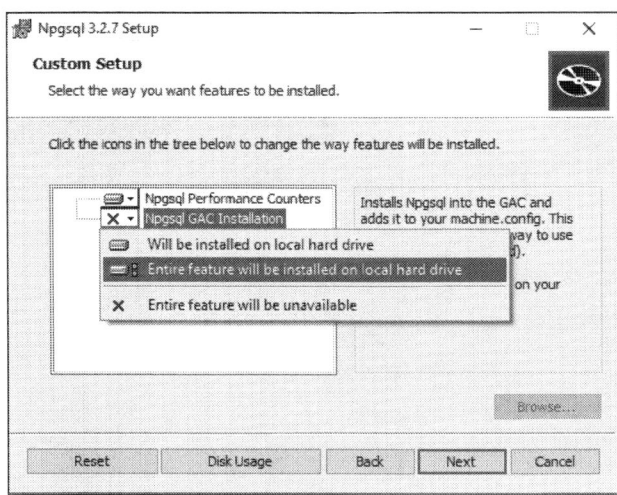

Figure 4.3: The Custom Setup page

The **Ready to install Npgsql 3.2.7** page appears.

11. Click the **Install** button, as shown in Figure 4.4.

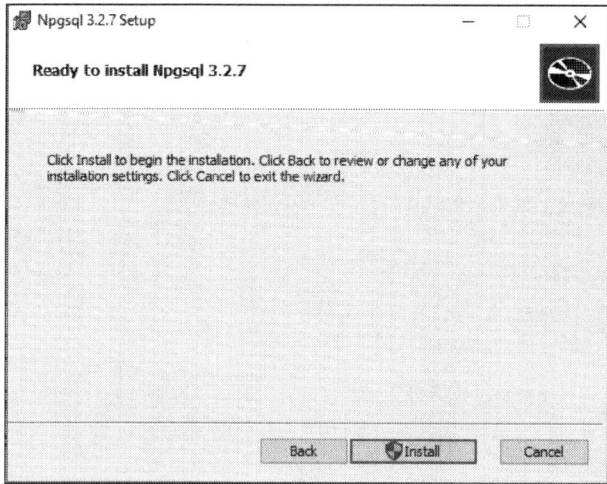

Figure 4.4: The Ready to install Npgsql 3.2.7 page

The **Installing Npgsql 3.2.7** page displays the progress of the Npgsql 3.2.7 installation, as shown in Figure 4.5.

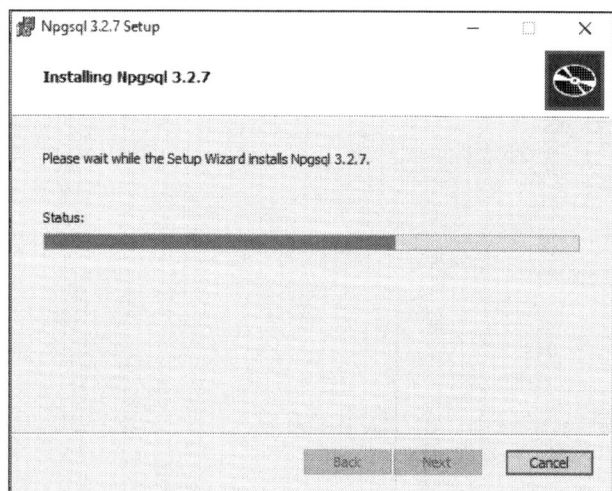

Figure 4.5: The Installing Npgsql 3.2.7 page

The **Completed the Npgsql 3.2.7 Setup Wizard** page appears.

12. Click the **Finish** button to exit the **Npgsql 3.2.7 Setup Wizard**, as shown in Figure 4.6.

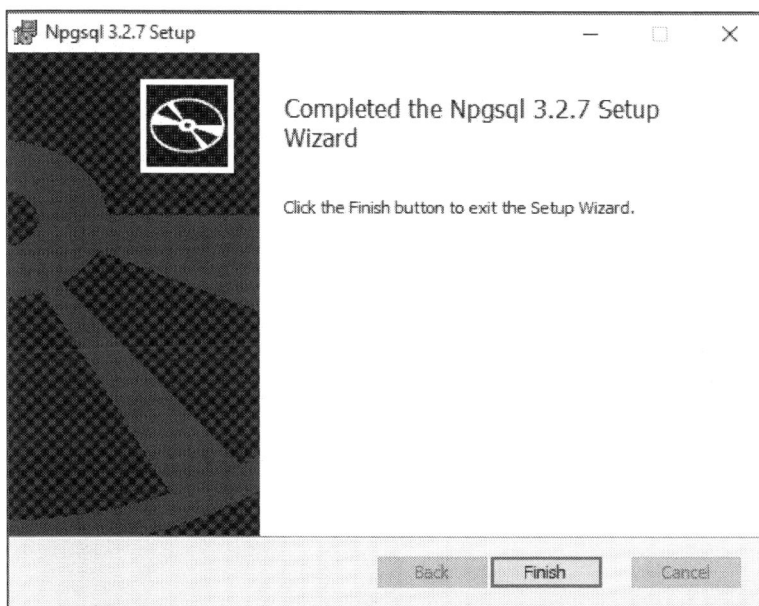

Figure 4.6: Finishing the Npgsql 3.2.7 Setup Wizard

After clicking the Finish button, the connector has been installed successfully.

Getting Data into Power BI

The first task in the process of creating a report is getting data from PostgreSQL into Power BI.

Perform the following steps to get data from the PostgreSQL database:

1. Launch Power BI Desktop.
2. Click the upper part of the **Get Data** button under the **External data** section of the **Home** tab, as shown in Figure 4.7.

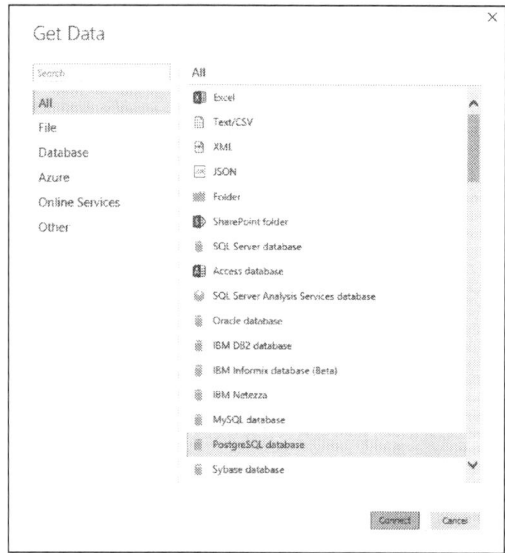

Figure 4.7: Clicking the upper part of the Get Data button

The **Get Data** dialog box appears.

3. Select the **All** option from the left pane.
4. Select the **PostgreSQL database** option from the right pane.
5. Click the **Connect** button, as shown in Figure 4.8.

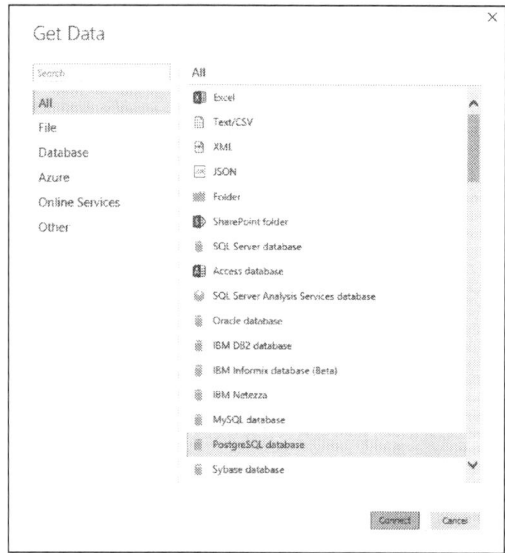

Figure 4.8: Connecting to PostgreSQL database

The **PostgreSQL database** dialog box appears.

6. Enter the name of the server in the **Server** text box.
7. Enter the name of the database in the **Database** text box.
8. Click the **OK** button, as shown in Figure 4.9.

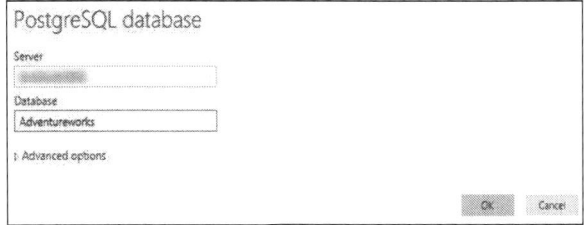

Figure 4.9: The PostgreSQL database dialog box

The **Navigator** dialog box appears.

9. From the **Display Options** section, select the tables that you want to import to Power BI Desktop. A view of the selected table appears in the right pane of the **Navigator** window.
10. Click the **Load** button, as shown in Figure 4.10.

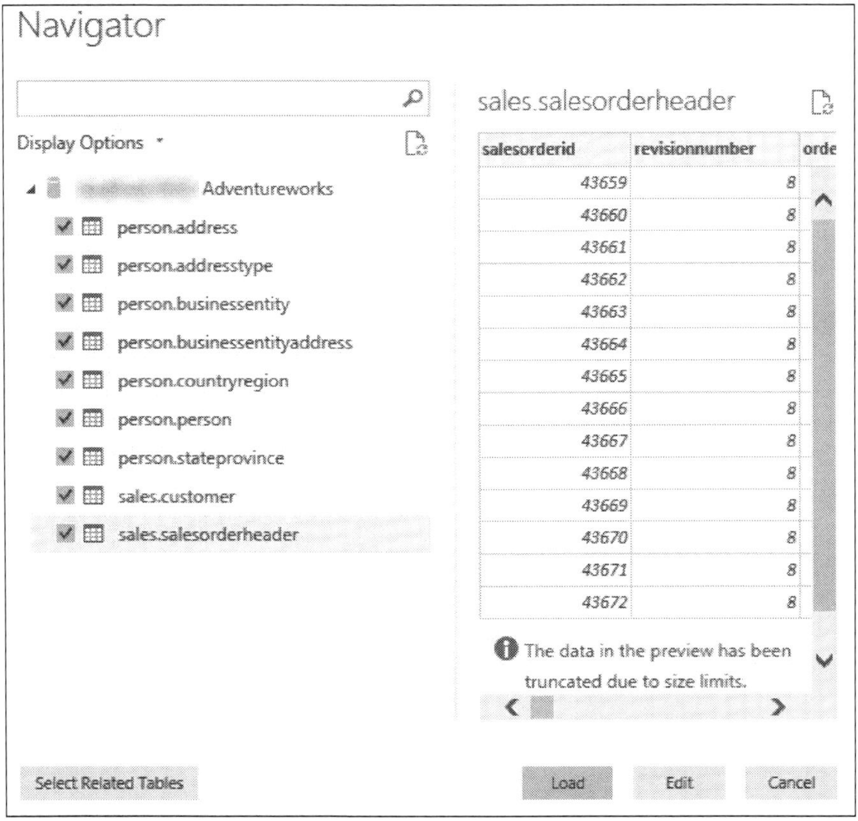

Navigator

sales.salesorderheader

Display Options

▲ ▣ Adventureworks
- ☑ person.address
- ☑ person.addresstype
- ☑ person.businessentity
- ☑ person.businessentityaddress
- ☑ person.countryregion
- ☑ person.person
- ☑ person.stateprovince
- ☑ sales.customer
- ☑ sales.salesorderheader

salesorderid	revisionnumber	orde
43659	8	
43660	8	
43661	8	
43662	8	
43663	8	
43664	8	
43665	8	
43666	8	
43667	8	
43668	8	
43669	8	
43670	8	
43671	8	
43672	8	

ⓘ The data in the preview has been truncated due to size limits.

Select Related Tables   Load   Edit   Cancel

Figure 4.10: Loading tables into Power BI

The selected tables load into Power BI Desktop and appear in the **FIELDS** pane, as shown in Figure 4.11.

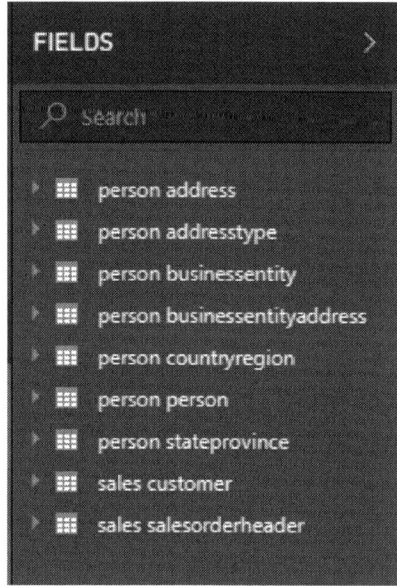

Figure 4.11: Displaying loaded tables

Data Modeling

After loading the data tables into Power BI, you need to shape data such that it can be used for creating intuitive reports. The way the data is shaped in Power BI is called data modeling. For example, if you have multiple tables from different data sources, you will need to create a relationship between these tables to use the data from these tables. You may also need to change the data type for a column, add or delete a new column or row, create a calculated column that returns a result based on data available in other columns, create a measure, and create a calculated table. All these operations fall within data modeling.

This section covers the following processes:
- Creating a relationship between tables
- DAX
- Creating a calculated column

Creating Relationship between Tables

The process of creating a relationship between tables is similar to the process discussed in Chapter 2. You can either use the Autodetect feature to automatically create a relationship between tables, or you can create it manually. You can also edit the relationship between tables as per your requirements. The relationship created between tables can be viewed from the **Relationships** view.

Using the Autodetect feature

Perform the following steps to create a relationship using the Autodetect feature:

1. Click the **Manage Relationships** button under the **Relationships** section of the **Home** tab, as shown in Figure 4.12.

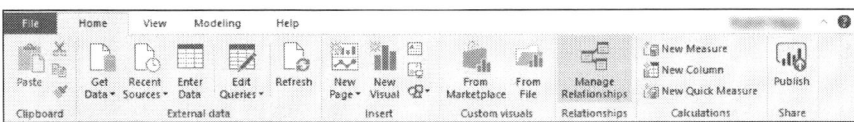

Figure 4.12: Clicking the Manage Relationships button

The **Manage relationships** window appears.

2. Click the **Autodetect** button to automatically detect relationships, as shown in Figure 4.13.

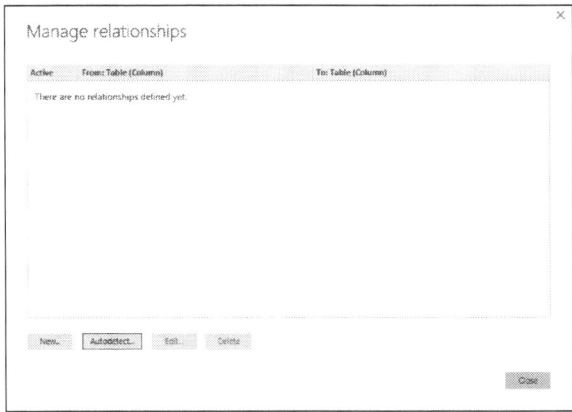

Figure 4.13: Using the Autodetect button

The **Detecting relationships** message box displays the progress of detecting relationships between tables. The **Autodetect** message box appears once the detection is done.

3. Click the **Close** button to close the **Autodetect** message box.

The **Manage relationships** window appears displaying the relationships between different tables, as shown in Figure 4.14.

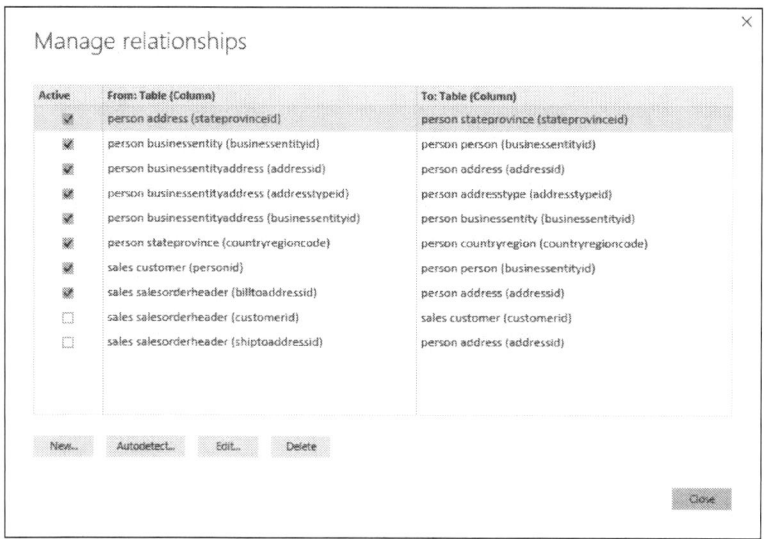

Figure 4.14:   The Manage relationships window with relationships

In the figure above, you can see that each successful relationship has active checkbox in the **Active** column.

4. Click the **Close** button to close the **Manage relationships** window.

You can see the relationships between tables in the **Relationships** view, as shown in Figure 4.15.

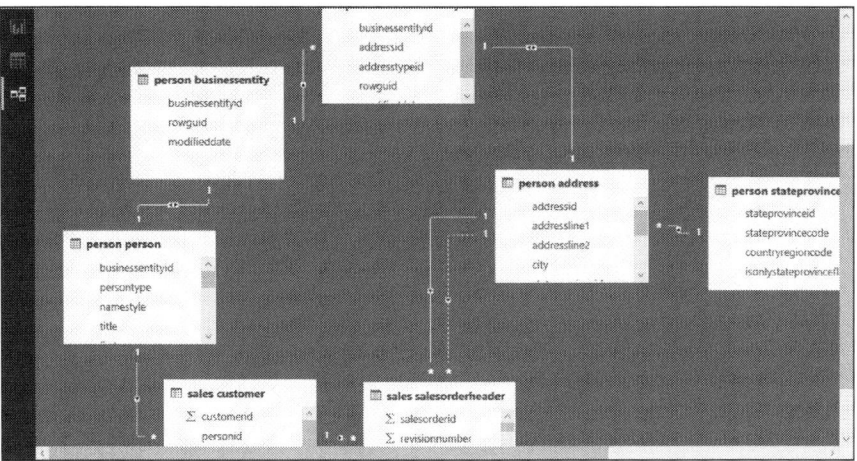

Figure 4.15: Viewing relationship in Relationships view

Creating Relationships Manually

Be aware that the Autodetect feature does not detect all relationships between all tables. Therefore, to create a relationship between tables that were missed by the Autodetect feature, you need to manually create the relationship. Perform the following steps to manually create a relationship:

1. Click the **Manage Relationships** button under the **Relationships** section of the **Home** tab. The **Manage relationships** window appears.
2. Click the **New** button to manually create relationships. The **Create relationship** dialog box appears.
3. Select the desired table from the first drop-down list. A list of columns associated with the selected table appears.
4. Select the column that you want to relate to another column.
5. Select another table from the second drop-down list. A list of columns associated with the selected table appears.
6. Select the column that you want to relate to the column selected in the first table.
7. Select the desired option from the **Cardinality** drop-down list to specify cardinality for the relationship.
8. Select the desired option from the **Cross filter direction** drop-down list.
9. Select the **Make this relationship active** checkbox to activate the relationship.
10. Click the **OK** button, as shown in Figure 4.16.

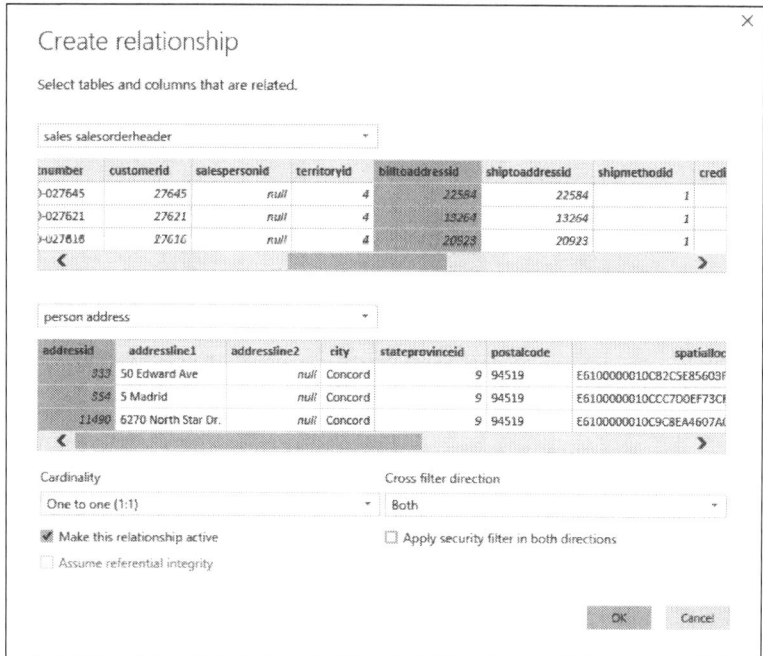

Figure 4.16: The Create relationship dialog box

The **Manage relationships** window appears and displays the new relationship.

11. Click the **Close** button to close the **Manage relationships** window.

12. Click the **Relationships** view to analyze the relationship between the selected tables.

Note

Cardinality is the degree of relationship, which specifies the number of occurrences of the first table linked to the number of occurrences of the second table.

Cross filter direction specifies the direction of filter applied to the tables of relationships.

Data Analysis Expressions

Data Analysis Expressions (DAX) is an expression (a set of functions, constants, and operators) written to apply calculations on data available in your model. The calculated data can be used for creating visuals. DAX expressions are commonly used to create calculated columns and calculated tables. There are three elements of DAX, including syntax, function, and context.

Note
A detailed description of the elements of DAX is covered in Chapter 2.

Creating a Calculated Column

A calculated column, as the name indicates, is a column that has a calculation applied to it. These calculations are applied in the form of DAX formulas. For example, if we have a table that contains employee details including first name, last name, address, phone number, and joining date, and we want the full name of each employee for our report, we would create a calculated column that contains the full name of each employee by using a DAX formula to join both strings. After creating the calculated column, we can use it in our report, just as we can with other columns. The created calculated column appears in the FIELDS pane similar to other columns. You can also assign the required name for the calculated column.

Perform the following steps to create a calculated column that concatenates the FirstName and LastName fields:
1. Click the **New Column** button under the **Calculations** section of the **Modeling** tab. A new column is added to the selected table in the **FIELDS** pane, and the formula bar appears above the Report canvas, as shown in Figure 4.17.

Figure 4.17: Displaying the formula bar

Note
The formula bar is the actual area where a DAX formula can be specified.

2. Enter the DAX formula in the formula bar, as shown in Figure 4.18.

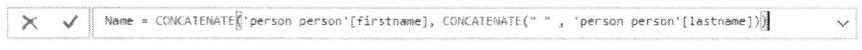

Figure 4.18: Using DAX formula

From the above figure, we can see the following:
    a. Name is the name of the calculated column.
    b. CONCATENATE is the name of the function that concatenates two strings.

      c. person person[firstname] states the firstname column in the person person table.

      d. person person[lastname] states the lastname column in the person person table.

3.   Click the **OK** (✓) icon to accept the changes.

After clicking the OK (✓) icon, the calculated column is created in the selected table in the **FIELDS** pane, as shown in Figure 4.19.

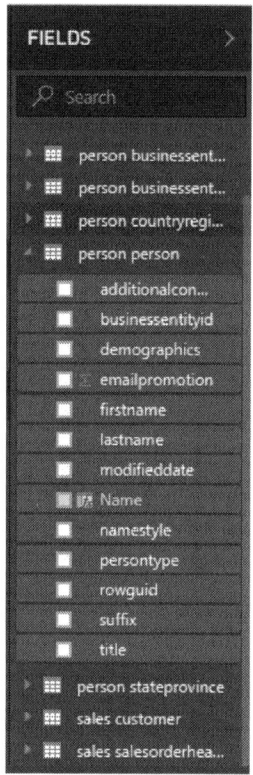

Figure 4.19: Displaying the calculated column

You can review the visual for the calculated column by selecting the desired visual from the **VISUALIZATIONS** pane and selecting fields to be displayed in the visual from the **FIELDS** pane, as shown in Figure 4.20.

firstname	lastname	Name
A.	Leonetti	A. Leonetti
A.	Wright	A. Wright
A. Scott	Wright	A. Scott Wright
Aaron	Adams	Aaron Adams
Aaron	Alexander	Aaron Alexander
Aaron	Allen	Aaron Allen
Aaron	Baker	Aaron Baker
Aaron	Bryant	Aaron Bryant
Aaron	Butler	Aaron Butler
Aaron	Campbell	Aaron Campbell
Aaron	Carter	Aaron Carter
Aaron	Chen	Aaron Chen
Aaron	Coleman	Aaron Coleman

Figure 4.20: Creating a visual for the calculated column

Creating a Report

Once you are done with the process of getting data from the PostgreSQL database and applying data modeling changes, you can create a report containing visuals in Power BI Desktop. We have created a report shown in Figure 4.21.

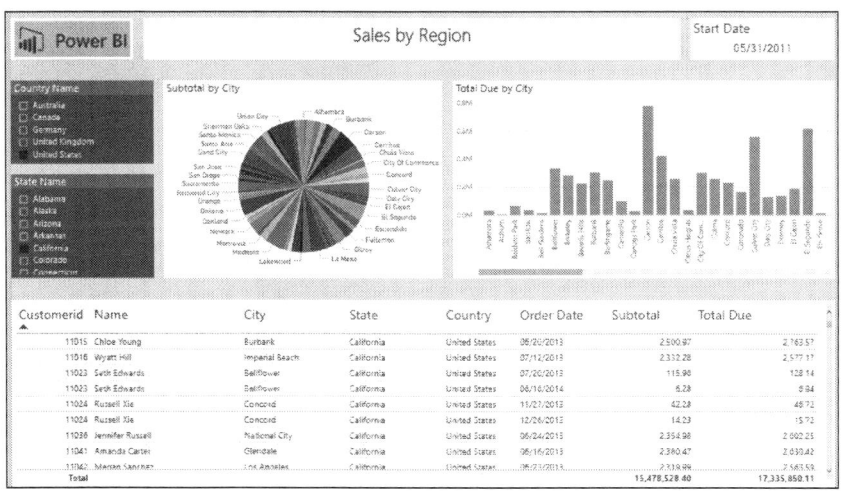

Figure 4.21: Creating a report

To create the above report, we performed the tasks listed below:

1. Set Report page details:
    a. Click the **Format** () icon.
    b. Expand the **Page Information** option. The related options appear.
    c. Enter the name of report page in the **Name** text box, as shown in Figure 4.22.

Figure 4.22: Setting related to the Page Information page

    d. Expand the **Page Size** option.
    e. Select the page size from the **Type** drop-down list.
    f. Expand the **Page Background** option.
    g. Click the **Color** drop-down button. The **Theme colors** palette appears.
    h. Select the color from the **Theme colors** palette.
    i. Set the **Transparency** option to **0%**, as shown in Figure 4.23.

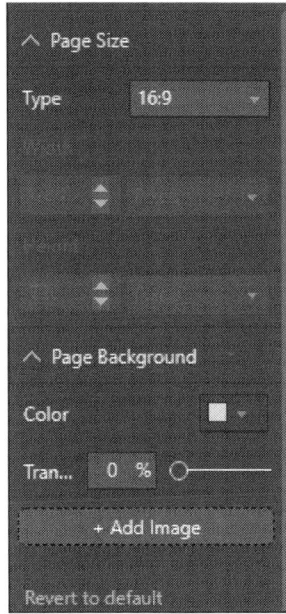

Figure 4.23: Setting report page details

2. Add and format an image (Power BI logo):
    a. Click the **Image** button under the **Insert** section of the **Home** tab. The **Open** dialog box appears.
    b. Navigate to the location where the image is saved.
    c. Select the image.
    d. Click the **Open** button. The selected image is inserted into the report page and the **FORMAT IMAGE** pane appears.
    e. Expand the **Scaling** option.
    f. Select the **Normal** option from the **Scaling** drop-down list.
    g. Drag the slider for the **Background** option to change its status to **On**.
    h. Expand the **Background** option to view settings related to the background.
    i. Click the **Color** drop-down button. The **Theme colors** palette appears.
    j. Select the color from the **Theme colors** palette.
    k. Set the **Transparency** option to **0%**, as shown in Figure 4.24.

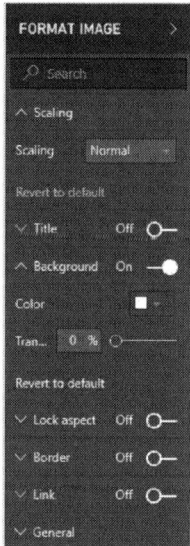

Figure 4.24: Setting scaling and background

l. Expand the **General** option. The related settings appear.

m. Set the value of width in the **Width** text box.

n. Set the value of height in the **Height** text box, as shown in Figure 4.25.

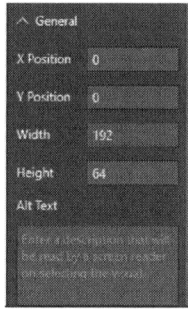

Figure 4.25: Specifying settings related to the General option

3. Add and format a text box:

a. Click the **Text box** button under the **Insert** section of the **Home** tab. A text box is inserted, and a formatting toolbar appears.

b. Enter the **"Sales by Region"** text.

c. Select the text.

d. Select the font from the **Font** drop-down list in the formatting toolbar.

e. Select the font size from the **Font Size** drop-down list in the formatting toolbar.

f. Click the **B** button to apply the bold style for the selected text.

g. Click the **Center** button to centralize the text, as shown in Figure 4.26.

Figure 4.26: Formatting toolbar

h. Drag the slider for the **Background** option to change its status to **On** in the **VISUALIZATIONS** pane.

i. Expand the **Background** option to view settings related to the background.

j. Click the **Color** drop-down button. The **Theme colors** palette appears.

k. Select the color from the **Theme colors** palette.

l. Set the **Transparency** option to **0%**.

m. Expand the **General** option. The related settings appear.

n. Set the values for the **X Position**, **Y Position**, **Width**, and **Height** text boxes, as shown in Figure 4.27.

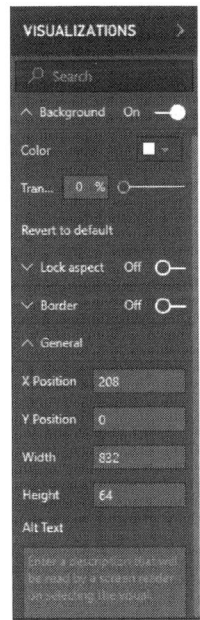

Figure 4.27: Specifying settings in the VISUALIZATIONS pane

4. Add and format a card visual:
   a. Click the **Card** visual from the **VISUALIZATIONS** pane. The selected visual appears in the report page.
   b. Drag the **orderdate** field from the **sales salesorderheader** table and drop it into the **Fields** value.
   c. Click the **Fields** drop-down button. A context menu appears.
   d. Select the **Earliest** option from the context menu. The card visual that you inserted into the report page starts displaying the earliest date.
   e. Click the **Format** icon to set the formatting settings.
   f. Expand the **Data label** option and specify the settings, as shown in Figure 4.28.

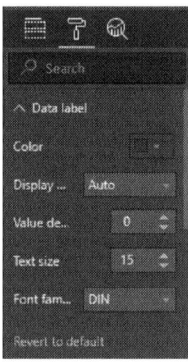

Figure 4.28: Specifying data label settings

   g. Drag the slider for the **Title** option to change its status to **On**.
   h. Expand the **Title** option and set the related settings, as shown in Figure 4.29.

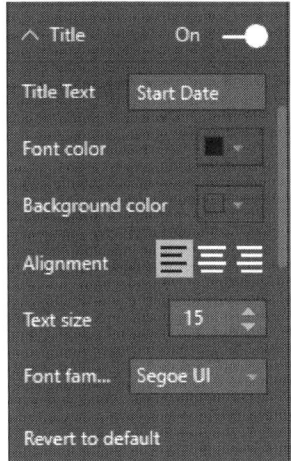

Figure 4.29: Specifying settings related to the Title option

    i.   Drag the slider for the **Background** option to change its status to **On**.

    j.   Set the color and transparency, as shown in Figure 4.30.

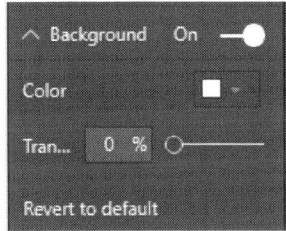

Figure 4.30: Specifying background settings

    k.   Expand the **General** option and set the related settings, as shown in Figure 4.31.

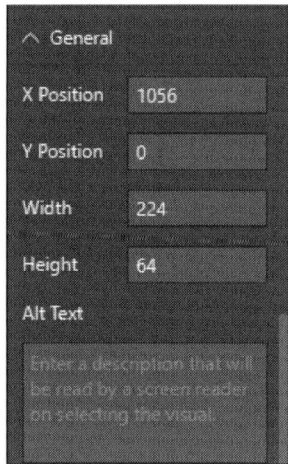

General

X Position	1056
Y Position	0
Width	224
Height	64

Alt Text

Enter a description that will be read by a screen reader on selecting the visual.

Figure 4.31: Specifying the General settings

5. Add the Slicer visual to display the country name:
   a. Click the **Slicer** visual from the **VISUALIZATIONS** pane. The selected visual appears in the report page.
   b. Drag the **name** field from the **person countryregion** table and drop it into the **Field** value.
   c. Drag the **countryregioncode** field from the **person countryregion** table and drop it into the **Report level filters** value.
   d. Select the **Basic filtering** option from the **Filter type** drop-down list.
   e. Select the checkboxes next to the country region codes. The name of the selected checkboxes will appear under the **Report level filters** section, as shown in Figure 4.32.

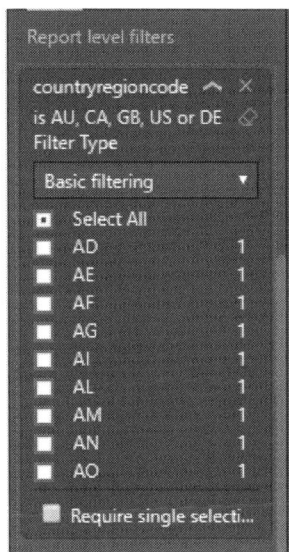

Figure 4.32: Applying basic filtering

    f.   Click the **Format** icon to set the formatting settings.

    g.   Expand the **General** option and specify the settings, as shown in Figure 4.33.

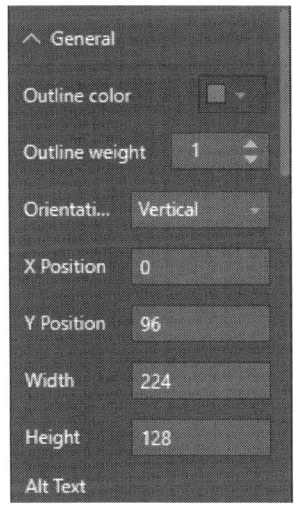

Figure 4.33: Specifying settings for the General option

h. Expand the **Items** option and specify the settings, as shown in Figure 4.34.

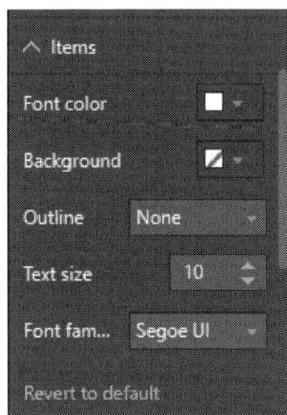

Figure 4.34: Specifying settings for the Items option

i. Drag the slider for the **Title** option to change its status to **On**.
j. Specify the settings related to the **Title** option, as shown in Figure 4.35.

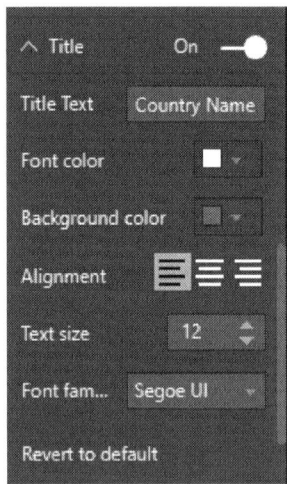

Figure 4.35: Specifying the settings related to the Title option

k. Drag the slider for the **Background** option to change its status to **On**.

l. Specify the settings related to the **Background** option, as shown in Figure 4.36.

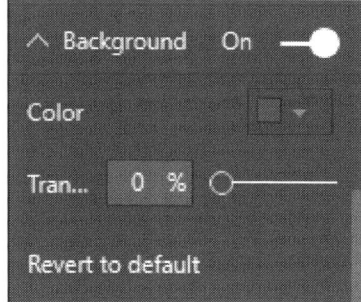

Figure 4.36: Specify settings related to the Background option

6. Add the Slicer visual to display the State name related to the country selected in the Country slicer:

a. Click the **Slicer** visual from the **VISUALIZATIONS** pane. The selected visual appears in the report page.

b. Drag the **name** field from the **person stateprovince** table and drop it into the **Field** value.

c. Click the **Format** icon to set the formatting settings.

d. Expand the **General** option and specify the settings, as shown in Figure 4.37.

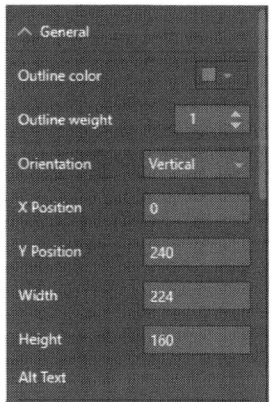

Figure 4.37: Specifying settings for the General option

e. Expand the **Items** option and specify the settings, as shown in Figure 4.38.

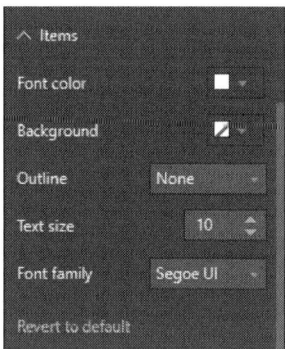

Figure 4.38: Specifying settings for the Items option

f. Drag the slider for the **Title** option to change its status to **On**.
g. Specify the settings related to the **Title** option, as shown in Figure 4.39.

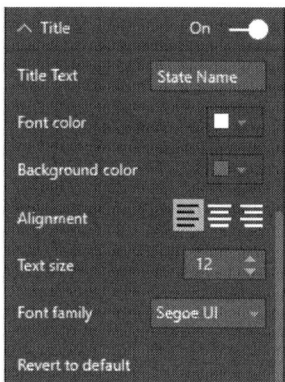

Figure 4.39: Specifying the settings related to the Title option

h. Drag the slider for the **Background** option to change its status to **On**.
i. Specify the settings related to the **Background** option, as shown in Figure 4.40.

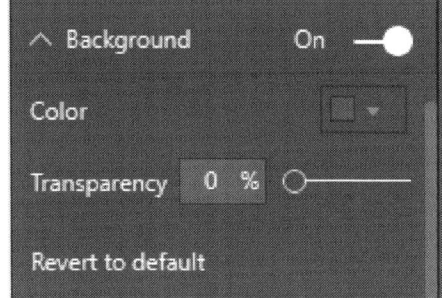

Figure 4.40: Specify settings related to the Background option

7. Add a Pie chart to visualize the subtotal filtered by city and format it:
   a. Click the **Pie chart** visual from the **VISUALIZATIONS** pane. The selected visual appears in the report page.
   b. Drag the **city** field from the **person address** table and drop it into the **Legend** value.
   c. Drag the **subtotal** field from the **sales salesorderheader** table and drop it into the **Values** value.
   d. Click the **Format** icon to set the formatting settings.
   e. Drag the slider for the **Title** option to change its status to **On**, expand it, and specify the settings, as shown in Figure 4.41.

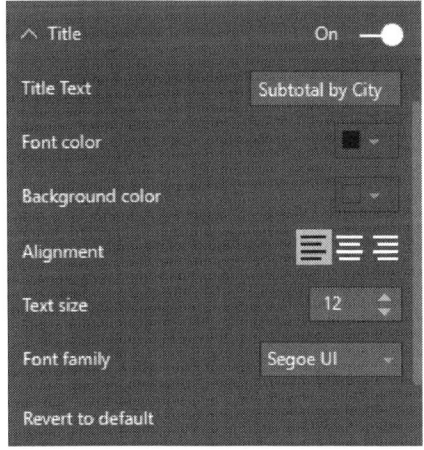

Figure 4.41: Specifying settings for the Title option

   f. Drag the slider for the **Background** option to change its status to **On**.

g. Specify the settings related to the **Background** option, as shown in Figure 4.42.

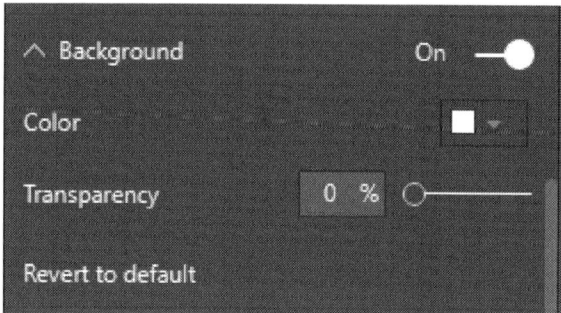

Figure 4.42: Specifying settings for the Background option

h. Expand the **General** option and specify settings related to it, as shown in Figure 4.43.

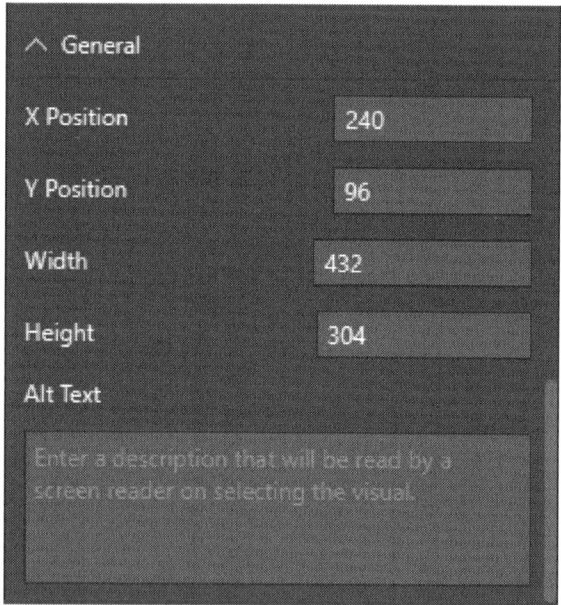

Figure 4.43: Specifying settings related to the General option

8. Add a Clustered column chart to visualize the total due filtered by city and format it:
   a. Click the **Clustered column chart** visual from the **VISUALIZATIONS** pane. The selected visual appears in the report page.
   b. Drag the **city** field from the **person address** table and drop it into the **Axis** value.
   c. Drag the **totaldue** field from the **sales salesorderheader** table and drop it into the **Value**.
   d. Click the **Format** icon to set the formatting settings.
   e. Expand the **General** option and specify settings related to it, as shown in Figure 4.44.

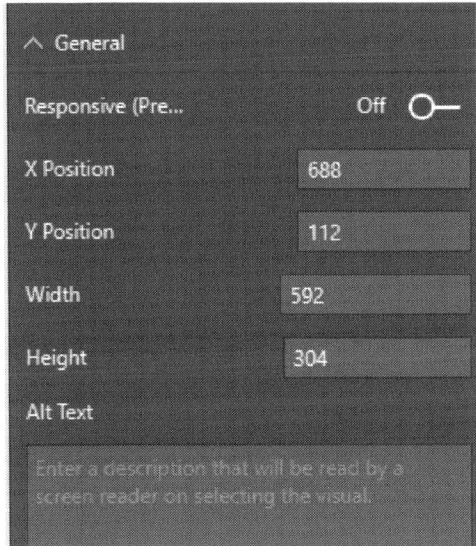

Figure 4.44: Specifying settings related to the General option

   f. Drag the slider for the **Title** option to change its status to **On**, expand it, and specify the settings, as shown in Figure 4.45.

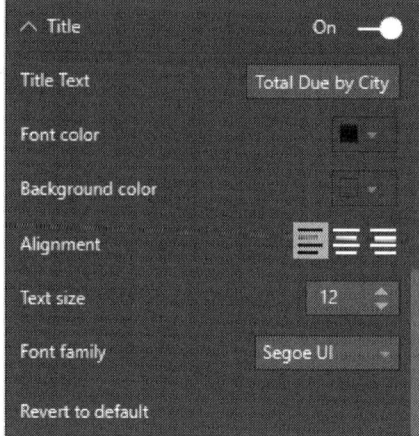

Figure 4.45: Specifying settings for the Title option

    g. Drag the slider for the **Background** option to change its status to **On**.

    h. Specify the settings related to the **Background** option, as shown in Figure 4.46.

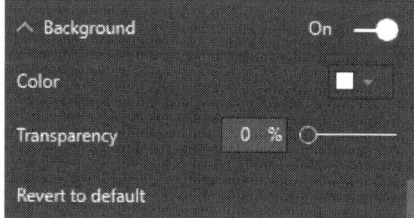

Figure 4.46: Specifying settings for the Background option

9. Add a table that displays the details of subtotal and total due per customer:

    a. Click the **Table** visual from the **VISUALIZATIONS** pane. The selected visual appears in the report page.

    b. Drag the **customerid** field from the **sales salesorderheader** table, drop it into the **Values** value, and rename **customerid** to **Customerid**.

    c. Drag the **Name** field from the **person** table and drop it into the **Values** value.

    d. Drag the **city** field from the **person address** table, drop it into the **Values** value, and rename **city** to **City**.

e. Drag the **name** field from the **person stateprovince** table, drop it into the **Values** value, and rename **name** to **State**.

f. Drag the **name** field from the **person countryregion** table, drop it into the **Values** value, and rename **name** to **Country**.

g. Drag the **orderdate** field from the **sales salesorderheader** table, drop it into the **Values** value, and rename **orderdate** to **Order date**.

h. Drag the **subtotal** field from the **sales salesorderheader** table, drop it into the **Values** value, and rename **subtotal** to **Subtotal**.

i. Drag the **totaldue** field from the **sales salesorderheader** table, drop it into the **Values** value, and rename **totaldue** to **Total Due**, as shown in Figure 4.47.

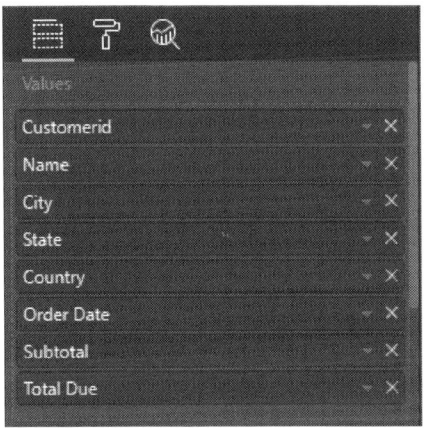

Figure 4.47: Adding values for the Table visual

j. Click the **Format** icon to set the formatting settings.

k. Expand the **General** option and specify settings related to it, as shown in Figure 4.48.

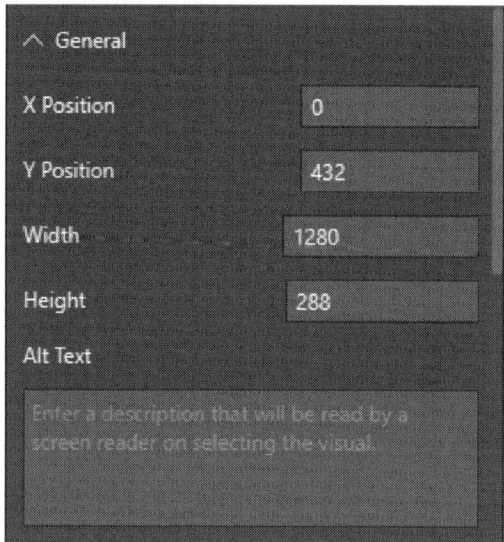

Figure 4.48: Specifying settings related to the General option

l. Expand the **Table style** option and select the **Minimal** option from the **Style** drop-down list, as shown in Figure 4.49.

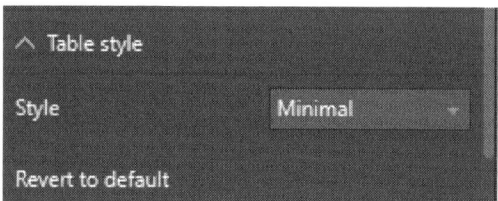

Figure 4.49: Setting the table style

m. Expand the **Grid** option and set the settings, as shown in Figure 4.50.

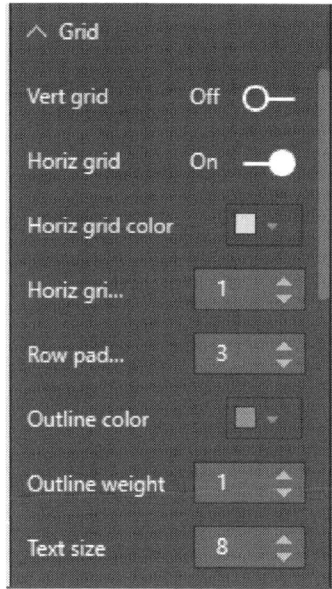

Figure 4.50: Specifying settings related to the Grid option

n. Expand the **Column headers** option and specify the settings related to it, as shown in Figure 4.51.

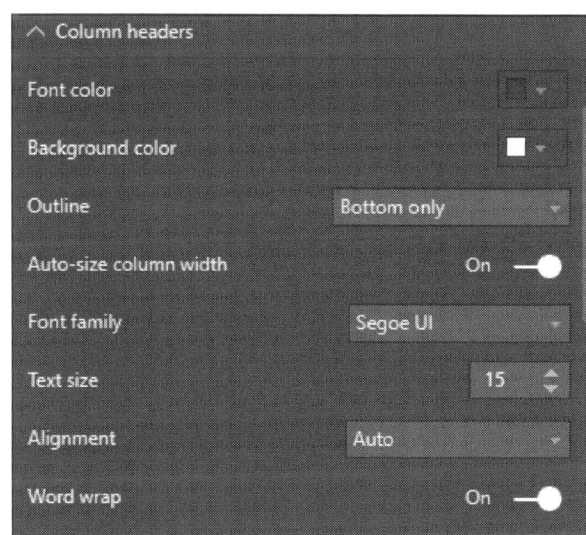

Figure 4.51: Settings related to the Column headers option

o. Expand the **Values** option and set the related settings, as shown in Figure 4.52.

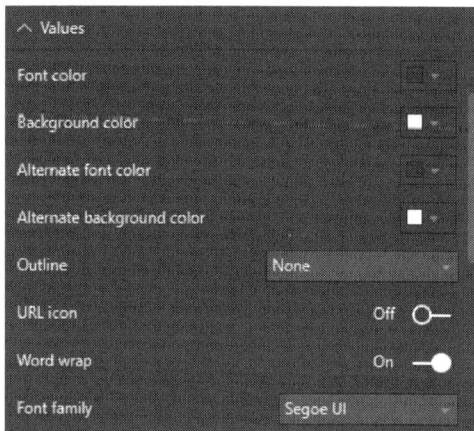

Figure 4.52: Specifying settings related to the Values option

p. Expand the **Total** option and specify the related settings, as shown in Figure 4.53.

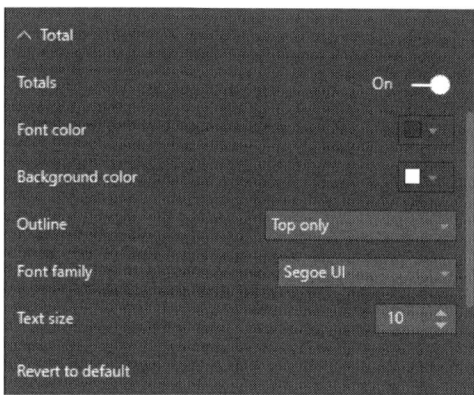

Figure 4.53: Specifying settings related to the Total option

q. Drag the slider to set the value of the **Background** option to **On**, expand it, and set the related settings, as shown in Figure 4.54.

Figure 4.54: Specifying settings related to the Background option

Saving the Report

It is a good practice to save a report regularly. Perform the following steps to save the report:

1. Select the **File** tab from the **Ribbon**. The Backstage View appears.
2. Select the **Save As** option from the Backstage View. The **Save As** dialog box appears.
3. Select the location wherein you want to save the report.
4. Type the desired name for the report in the **File name** text box.
5. Click the **Save** button, as shown in Figure 4.55.

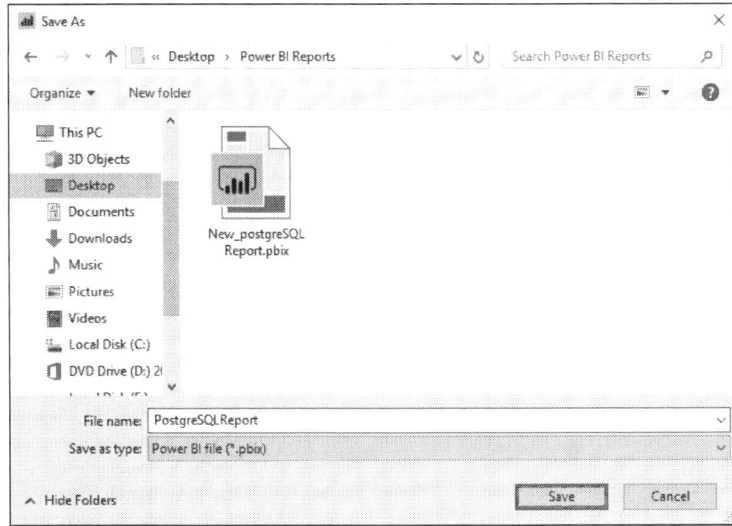

Figure 4.55: Saving a report

The report is saved with the specified name.

Publishing the Report

You can publish a report created in Power BI Desktop to Power BI Service and make the report accessible to others.

Perform the following steps to publish a report:
1. Open the report in Power BI Desktop.
2. Click the **Publish** button under the **Share** section of the **Home** tab, as shown in Figure 4.56.

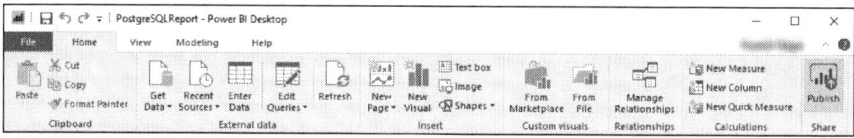

Figure 4.56: Publishing a report

The **Publish to Power BI** dialog box appears.
3. Select the desired destination from the **Select a destination** text area.
4. Click the **Select** button, as shown in Figure 4.57.

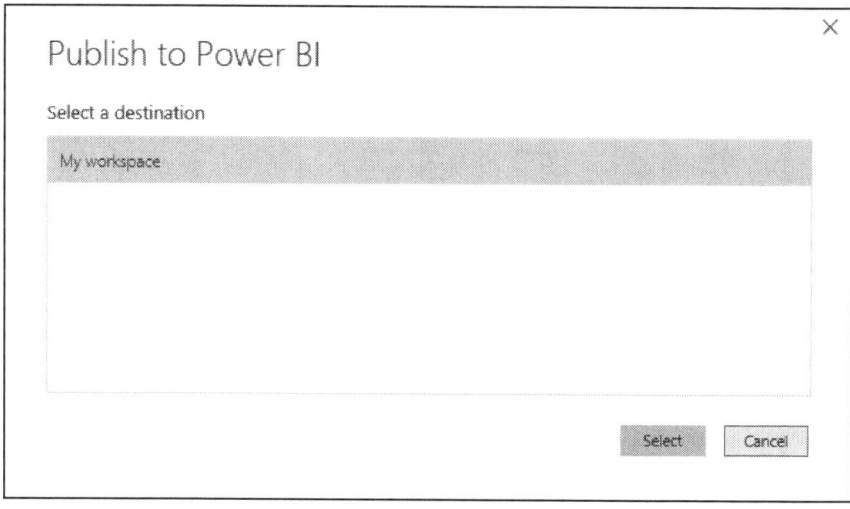

Figure 4.57: The Publish to Power BI dialog box

The **Publishing to Power BI** dialog box displays the status of publishing the report to Power BI, as shown in Figure 4.58.

Figure 4.58: Displaying the status of publishing the report

Once publishing is successful, the **Publishing to Power BI** dialog box displays the **Success** message. You can open the report in Power BI by clicking the **Open 'PostgreSQLReport.pbix' in Power BI** link.

5. Click the **Got it** button, as shown in Figure 4.59.

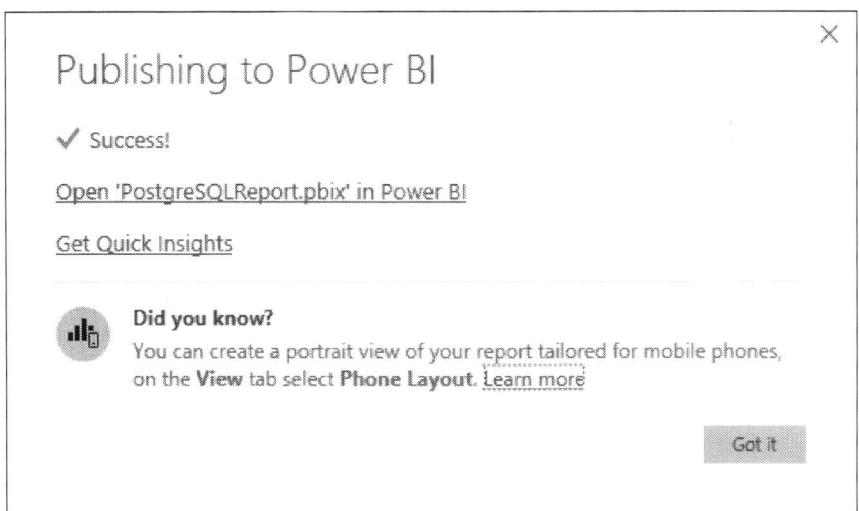

Figure 4.59: The Publishing to Power BI dialog box

The report is published successfully to Power BI Service, as shown in Figure 4.60.

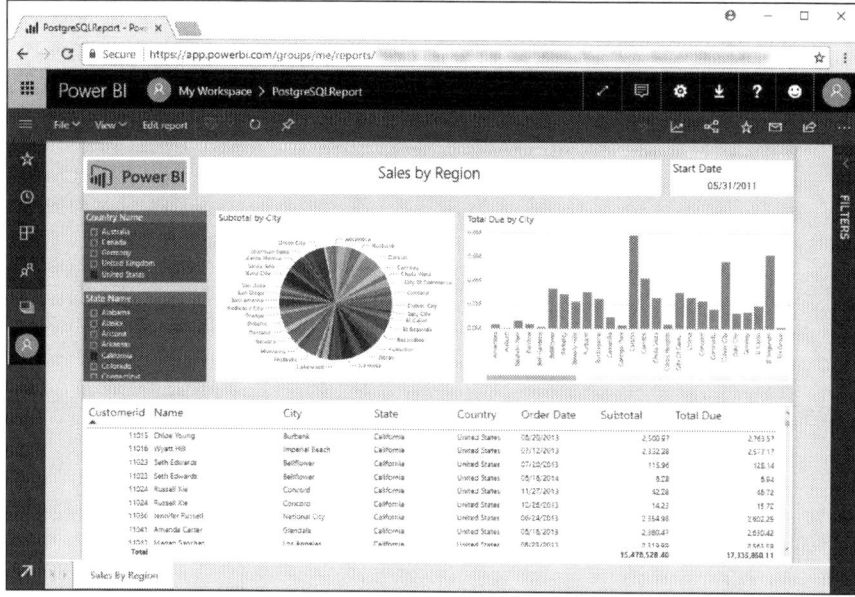

Figure 4.60: Displaying the published report

Data Refresh in Power BI

Displaying the most up to date information helps users in making better business decisions. Therefore, you should always use accurate and updated data for creating reports in Power BI. This can be done by creating a report, publishing it to Power BI Service, and using the data refresh functionality of Power BI to update data in the reports. You can also set a scheduled refresh in Power BI to automatically update visuals when there is an update in data. You can also manually refresh the reports by clicking the Refresh button in Power BI.

In Power BI Service, a dataset is a subset of data taken from the data source and includes information about the data sources and their credentials. A dataset is created automatically and appears under the Datasets tab in Power BI Service.

Setting Up Scheduled Refresh

You can apply a scheduled refresh by either using the **Schedule refresh** icon next to the dataset name, or by using the **Settings** option from the menu that appears when you click the **Ellipsis** icon (...) next to the dataset name. For successful configuration of a scheduled refresh, you need to set the following settings:
- Gateway connection
- Data source credentials
- Schedule refresh

Gateway Setup

A gateway is piece of software that allows users to access data located on an on-premises system or network so that it can be used in a cloud service later. Only authorized users are permitted to access the gateway. Similar to a gatekeeper who allows only authorized personnel to gain entry, the gateway attends all connection requests but grants access to only those users who meet certain criteria.

Power BI provides the following two types of gateways:
- **On-premises data gateway (personal mode):** Allows only one person to connect to data sources, does not allow sharing of reports, and it can only be used by Power BI. This gateway is for personal use wherein the user installs the gateway on his/her computer and the data source is located on-premises.
- **On-premises data gateway:** Can be used and shared by multiple users. This type of gateway can be used by multiple services including Power BI, PowerApps, and Azure Logic Apps, etc. It supports schedule refresh as well as DirectQuery for Power BI.

You need to install a data gateway on the machine running PostgreSQL and configure it such that it can be used by Power BI.

Installing On-Premises Data Gateway

Perform the following steps to download and install an on-premises data gateway:
1. Browse to the following link:
   https://app.powerbi.com/
2. Sign in to Power BI Service with the same credentials used in Power BI Desktop.
3. Click the **Download** icon. A list of options appears.
4. Click the **Data Gateway** option, as shown in Figure 4.61.

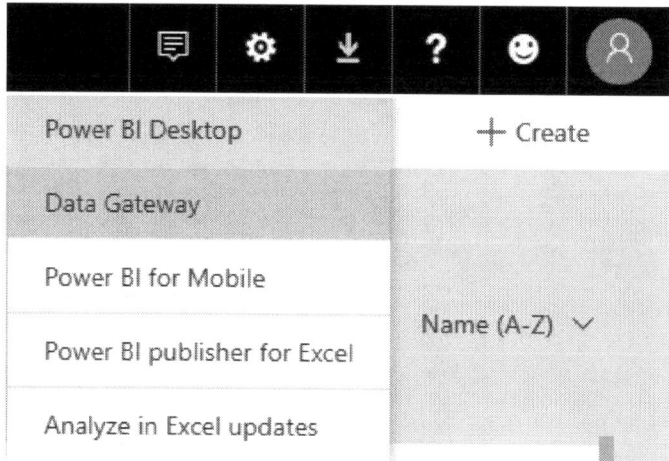

Figure 4.61: Clicking the Data Gateway option

The Power BI page displays a link to download Gateway.

5. Click the **DOWNLOAD GATEWAY** button, as shown in Figure 4.62.

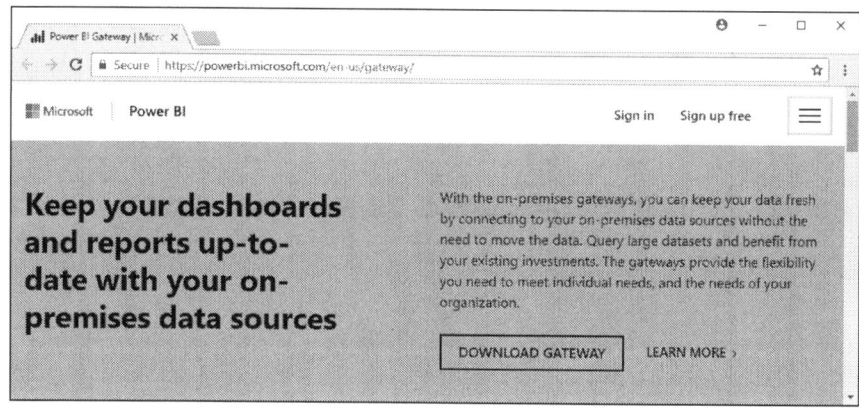

Figure 4.62: Download page

Once the download is complete, the **PowerBIGatewayInstaller.exe** file will display in the **Downloads** folder.

6. Double-click the **PowerBIGatewayInstaller.exe** in the **Downloads** folder. The **On-premises data gateway installer** wizard appears.

7. Click the **Next** button to start data gateway installation, as shown in Figure 4.63.

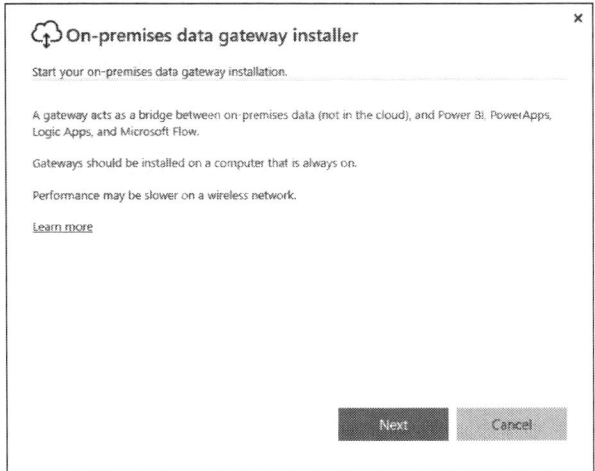

Figure 4.63: Starting the installation

The **Choose the type of gateway you need** page of the **On-premises data gateway installer** wizard appears.

8. Select the **On-premises data gateway (recommended)** radio button to install the on-premises data gateway.

9. Click the **Next** button, as shown in Figure 4.64.

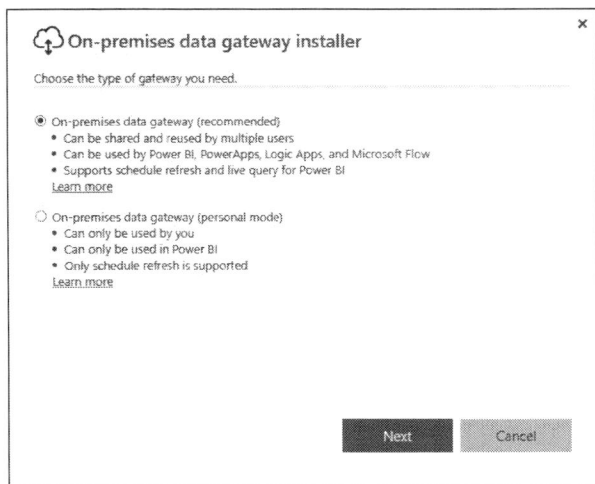

Figure 4.64: Selecting the gateway type

The **Getting ready to install the on-premises data gateway** page appears, which displays the progress of collecting the required information and files to install the on-premises data gateway, as shown in Figure 4.65.

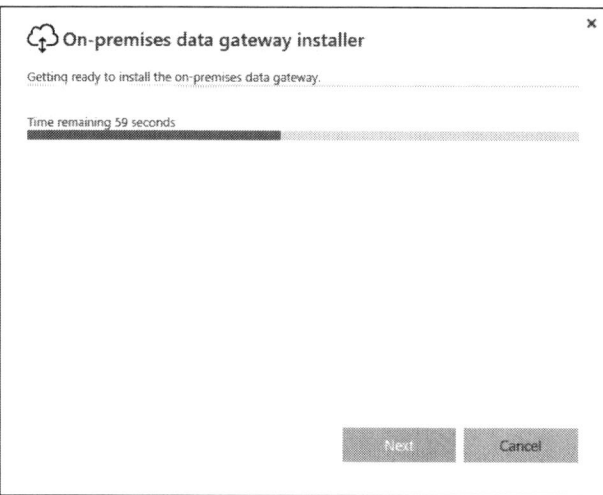

Figure 4.65: The Getting ready to install the on-premises data gateway page

The **Reminder before you install** page appears. This page gives warning for using the gateway.

10. Click the **Next** button, as shown in Figure 4.66.

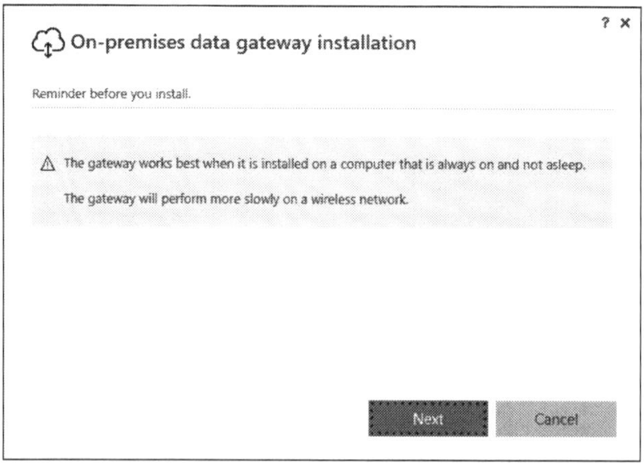

Figure 4.66: Reminder before installation

The **Getting ready to install the on-premises data gateway** page appears.

11. Specify the directory path wherein you want to install the gateway in the **Install to** text box.
12. Click the **I accept the terms of use and privacy statement** checkbox.
13. Click the **Install** button, as shown in Figure 4.67.

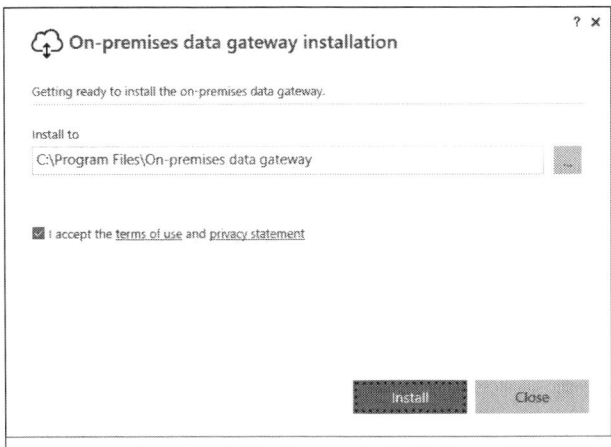

Figure 4.67: Specifying installation folder

The **Installing your on-premises data gateway** page displays the progress of installing the data gateway, as shown in Figure 4.68.

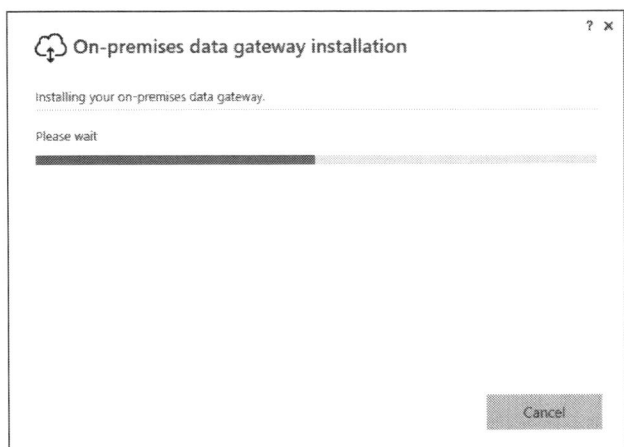

Figure 4.68: Progress of installation of the data gateway

The **Almost done** page of the **On-premises data gateway installation** wizard appears stating that installation was successful. This page also states that you need to sign in to register your gateway.

Configuring Gateway

Once the installation is successful, you need to configure the gateway so that it can be used with Power BI. For configuring the gateway, you need to continue with the steps followed in the earlier section, "Downloading and Installing the On-premises Data Gateway." Perform the following steps to configure/register gateway:

1. Enter the email address with which you want to register your gateway in the **Email address to use with this gateway** text box of the **On-premises data gateway** window.
2. Click the **Sign in** button, as shown in Figure 4.69.

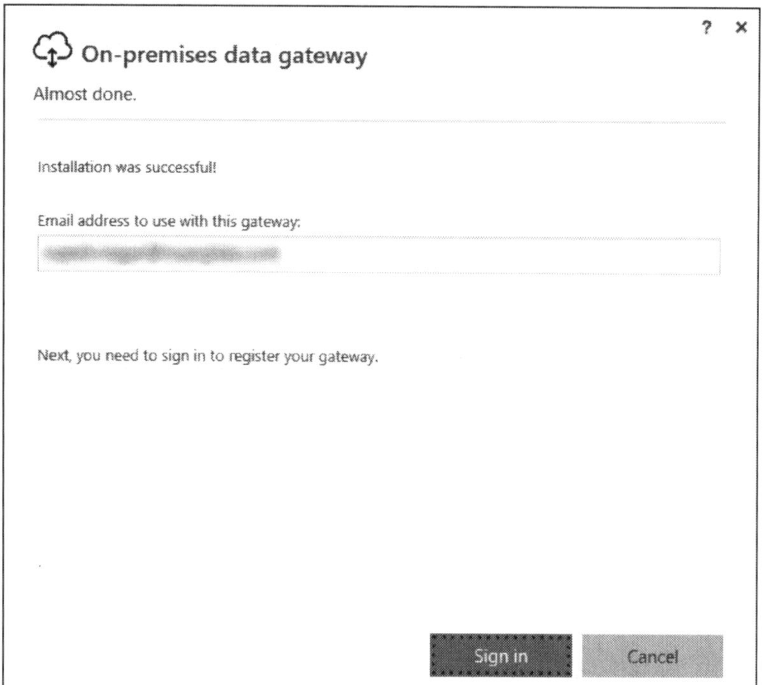

Figure 4.69: Specifying the email address

The **Sign in** window appears.

3. Enter the related password in the **Password** text box.
4. Click the **Sign in** button.

The next page of the **On-premises data gateway** window appears.

5. Select the **Register a new gateway on this computer** radio button to register a new gateway.
6. Click the **Next** button, as shown in Figure 4.70.

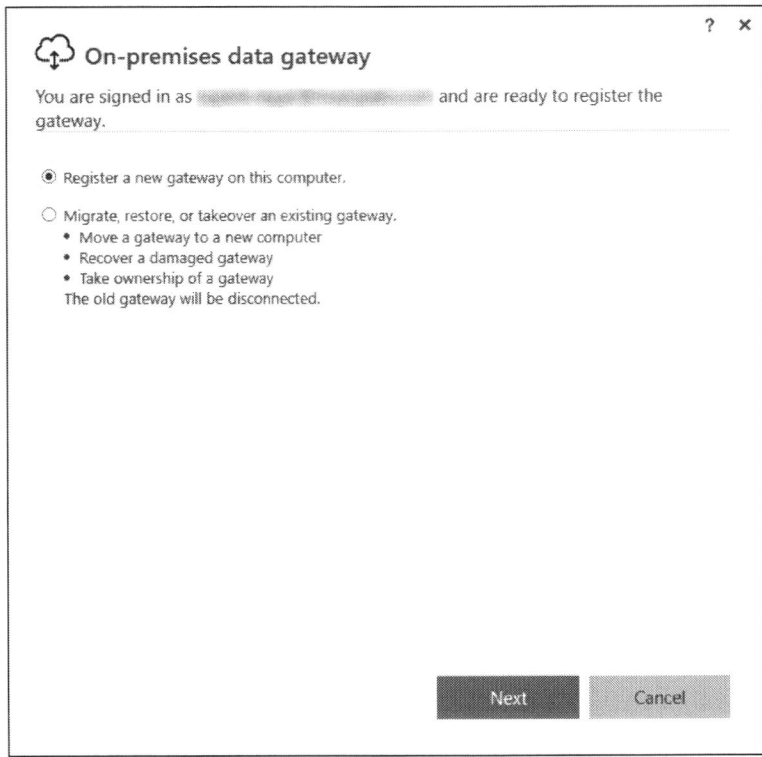

Figure 4.70: Specifying the type of gateway registration

The next page of the **On-premises data gateway** window asks you to specify a name and recovery key for the gateway.

7. Enter the name of a gateway in the **New on-premises data gateway name** text box.
8. Enter the desired recovery key for the gateway in the **Recovery key** text box.
9. Enter the same recovery key in the **Confirm recovery key** text box.
10. Click the **Configure** button, as shown in Figure 4.71.

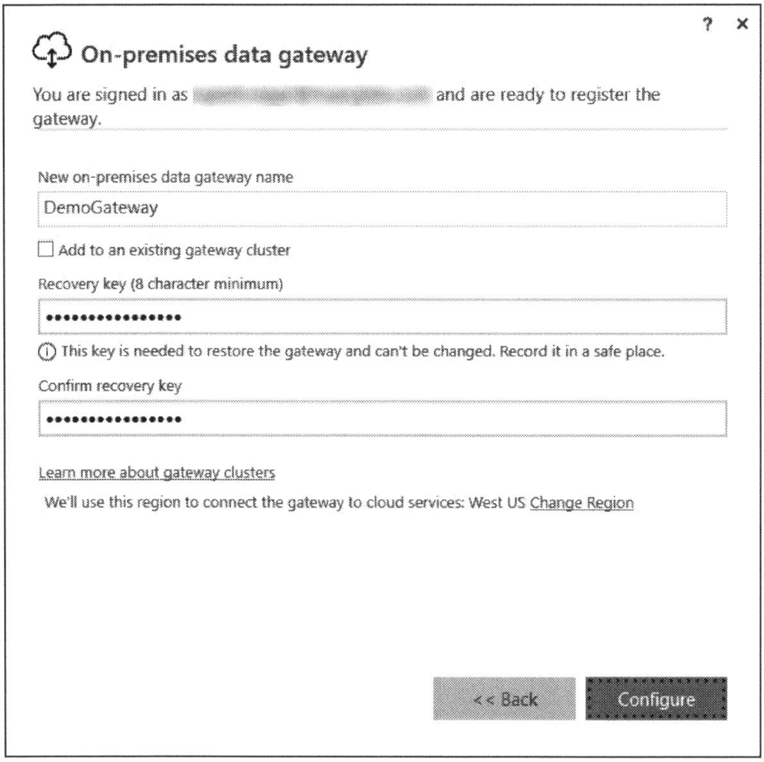

Figure 4.71: Configuring on-premises data gateway

Once the gateway is configured, you will see the following status message: **"The gateway DemoGateway is online and ready to be used."**

11. Click the **Close** button to close the **On-premises data gateway** window, as shown in Figure 4.72.

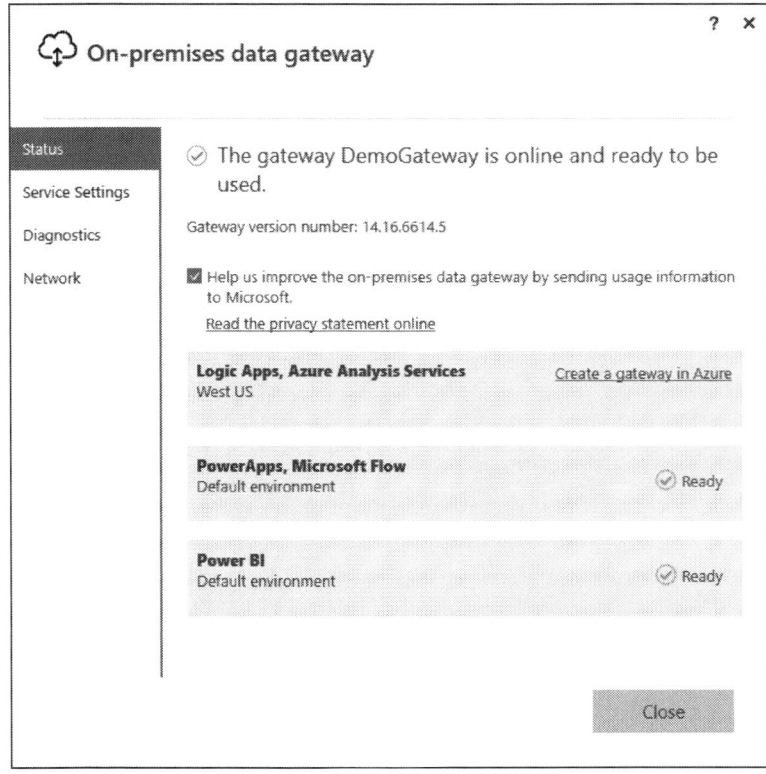

Figure 4.72: Completion status of gateway configuration

Adding Data Source

After installing the on-premises data gateway, you can add a data source (Post-greSQL Server) to be used with the gateway. You can find the list of available gateways under the GATEWAY CLUSTERS section of the Gateways window. The settings related to the selected gateway cluster appear in the right pane of the Gateways window.

Perform the following steps to add a data source to the gateway created earlier:

1. Launch Power BI Service.
2. Click the **Settings** icon. A list of options appears.
3. Click the **Manage gateways** option, as shown in Figure 4.73.

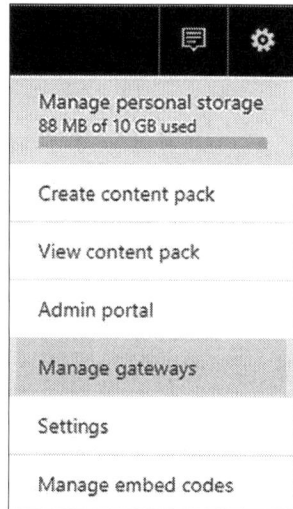

Figure 4.73: Clicking the Manage gateways option

The **Gateways** window appears. In the left pane, you will see a list of available gateways under the **GATEWAY CLUSTERS** section. The right pane displays information about the selected gateway and other settings related to the gateway under the **Gateway Cluster Settings** tab, as shown in Figure 4.74.

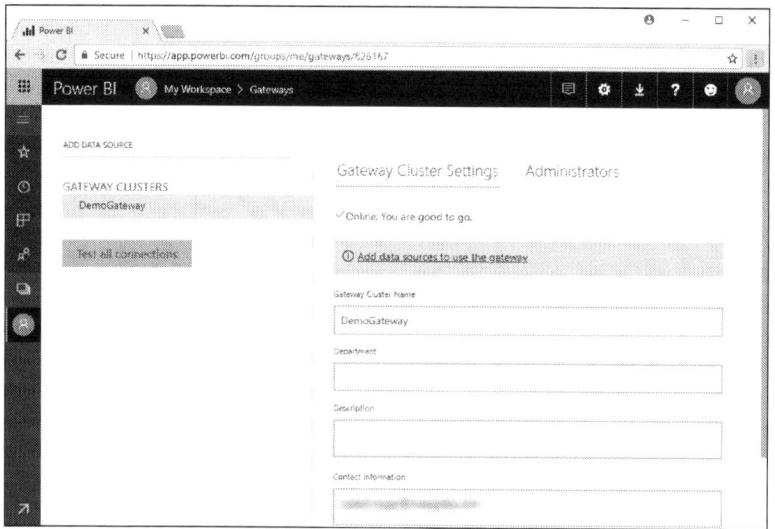

Figure 4.74: Gateway Cluster Settings

4. Click either **Add data sources to use this gateway** link or select the **Ellipsis** icon (...) next to the gateway and click the **ADD DATA SOURCE** option from the menu, as shown in Figure 4.75.

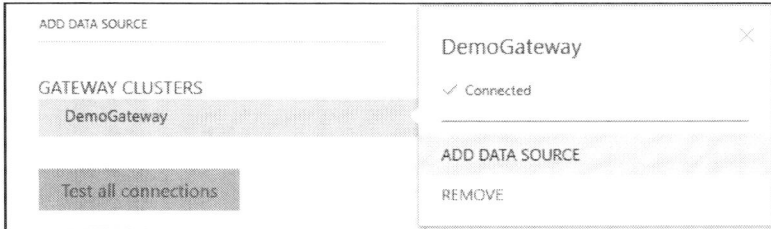

Figure 4.75: Clicking the ADD DATA SOURCE option

The **Data Source Settings** pane appears.

5. Enter the desired name for the data source in the **Data Source Name** text box.
6. Select the **PostgreSQL** option from the **Data Source Type** drop-down list to connect the created gateway to the PostgreSQL data source.
7. Enter the name of PostgreSQL server in the **Server** text box.
8. Enter the same name of the database that you used earlier, in the **Database** text box, as shown in Figure 4.76.

Data Source Settings

Data Source Name

DemoDataSource

Data Source Type

PostgreSQL ▼

Server

Database

Figure 4.76: Specifying data source settings

9. Enter the desired username in the **Username** text box.
10. Enter the desired password in the **Password** text box.
11. Expand the **Advanced settings** option. The related settings appear.
12. Select the desired privacy level setting from the **Privacy Level setting for this data source** drop-down list.
13. Click the **Add** button, as shown in Figure 4.77.

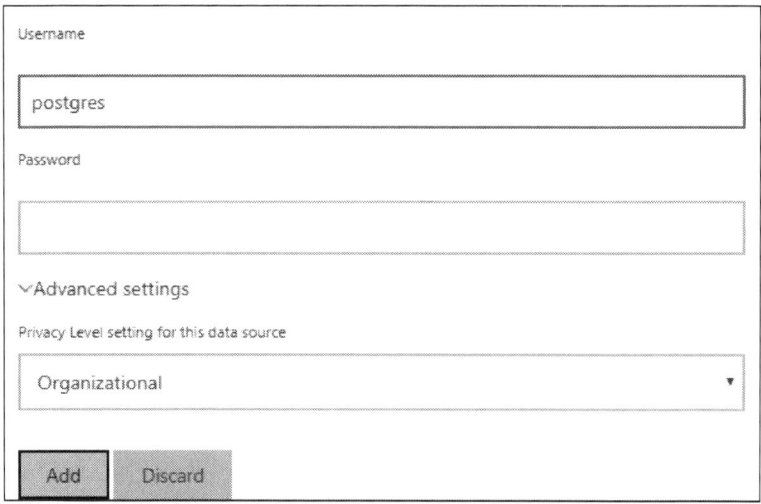

Figure 4.77: Specifying username and password

After clicking the **Add** button, the **Connection Successful** message appears, which states that the connection has been established, as shown in Figure 4.78.

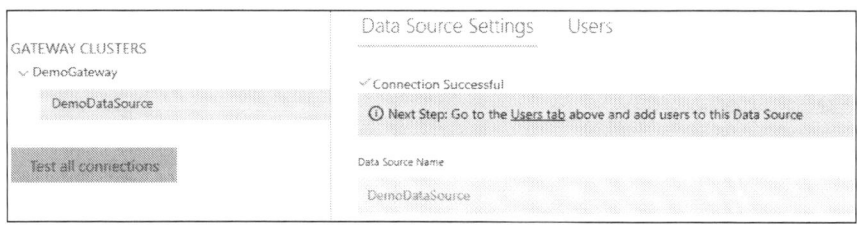

Figure 4.78: Connection successful

14. Select the **Users** tab to add users to the created data source.
15. Specify the email address of the person that you want to allow to publish reports using this data source.

16. Click the **Add** button. The specified person is added to the list box, as shown in Figure 4.79.

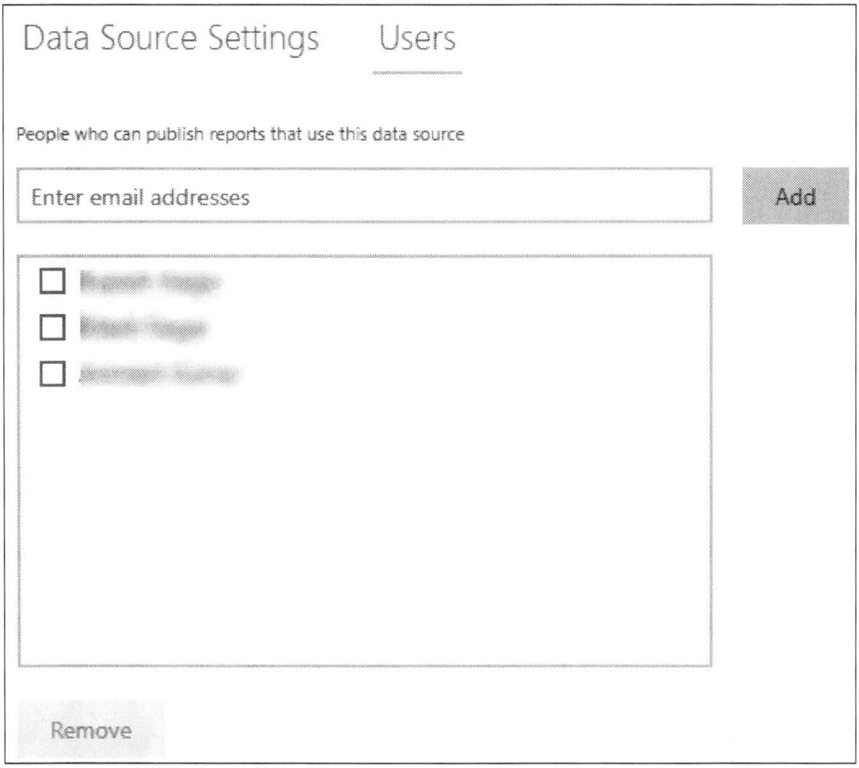

Figure 4.79: Adding users

Configuring Scheduled Refresh

As discussed earlier, you can configure a scheduled refresh for a dataset only after completing the setup of the data gateway.

Perform the following steps to configure a scheduled refresh for a dataset:

1. Launch Power BI Service.
2. Select the **My Workspace** option from the left pane.
3. Select the **Datasets** tab in the right pane. A list of available datasets appears.
4. Click the **Schedule refresh** icon beside the desired dataset, as shown in Figure 4.80.

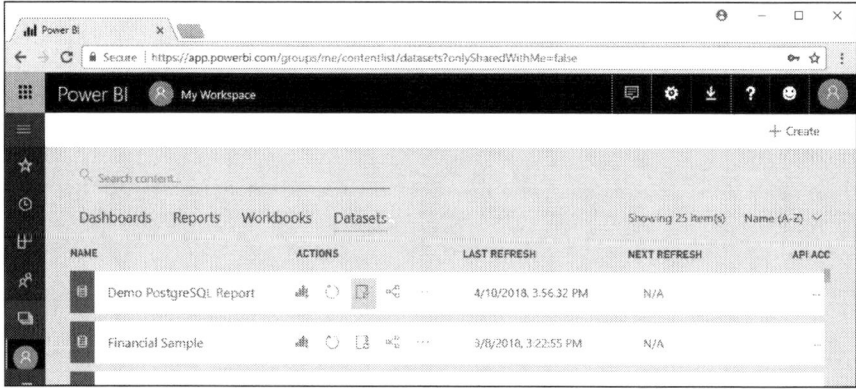

Figure 4.80: Clicking the Schedule refresh icon

The **Settings** page for the selected dataset appears.

5. Expand the **Gateway connection** node to expand the settings related to the gateway connection. The related settings appear.

6. Select the **Use an on-prem data gateway** radio button. You will see the status as online.

7. Click the **Apply** button to apply the changes, as shown in Figure 4.81.

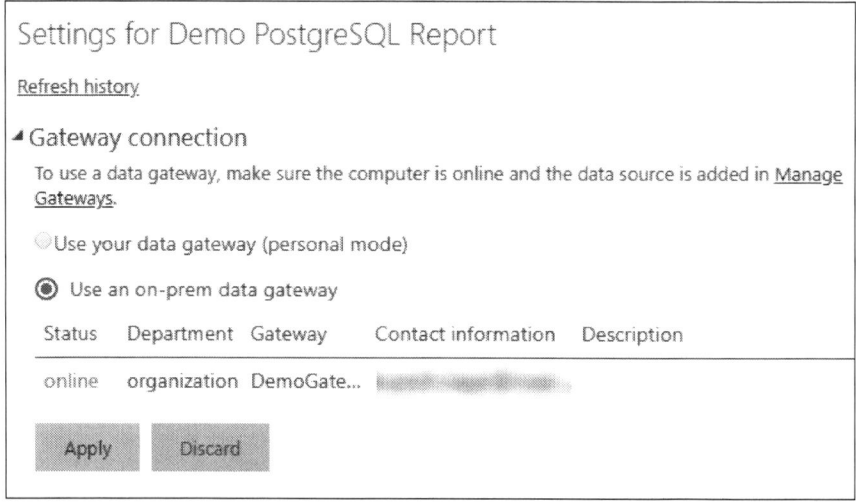

Figure 4.81: Setting Gateway connection

8. Expand the **Scheduled refresh** node and specify settings related to schedule refresh, including the frequency and time slots to refresh the dataset.
9. Drag the slider under the **Keep your data up to date** section to change its status to **On**.
10. Select the desired frequency for a refresh from the **Refresh frequency** drop-down list.
11. Select the desired time zone from the **Time zone** drop-down list.
12. Click the **Send refresh failure notification email to me** checkbox to apply a setting such that you receive an email if the refresh fails.
13. Click the **Apply** button, as shown in Figure 4.82.

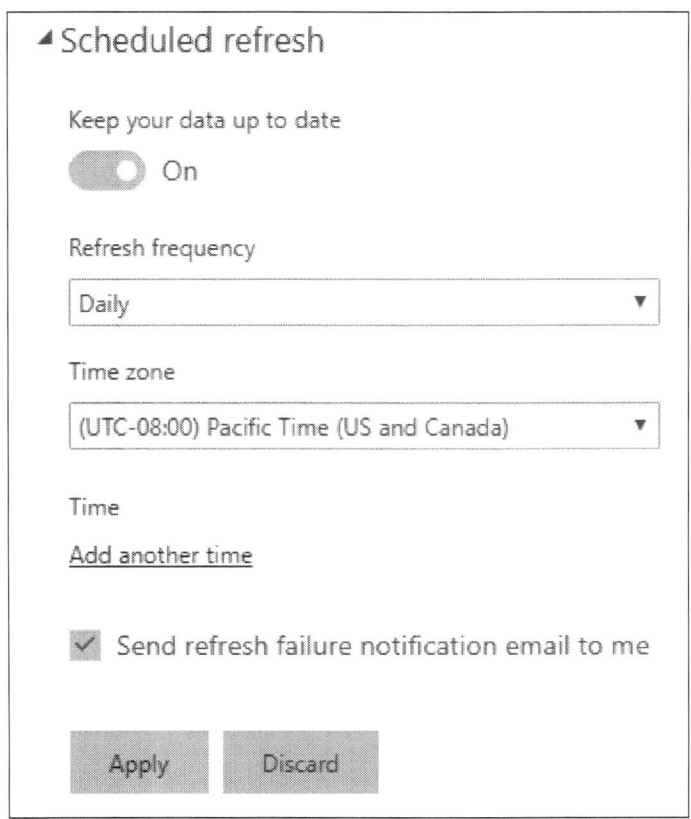

Figure 4.82: Configuring scheduled refresh

After clicking the Apply button, the scheduled refresh is configured.

Creating Content Packs

A content pack is a complete package of your dashboard, report, and dataset that can be shared with other users in your organization. You can create the content pack and publish it to the team. As you publish the content pack, it becomes available in a centralized repository called **AppSource**. This repository helps team members to easily locate reports and datasets published for them.

You can locate the content packs in a central repository only when you are a member of a particular group, such as the entire organization, distribution group, security group, or Office 365 group, to which the content pack is published.

Note
You will need a Power BI Pro account for creating and accessing an organizational content pack.

Perform the following steps to create and publish a content pack:
1. Launch Power BI Service.
2. Click the **Settings** icon. A drop-down menu appears.
3. Click the **Create content pack** button in the drop-down menu, as shown in Figure 4.83.

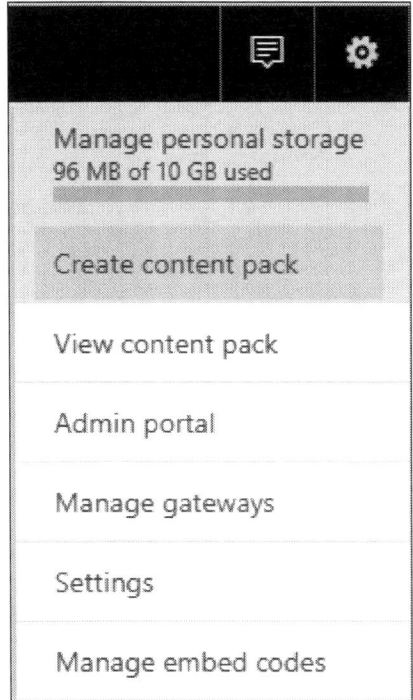

Figure 4.83: Creating a content pack

The **Create content pack** page appears.

4. Select either the **Specific groups** radio button to allow a specific group to access this content pack, or select the **My entire organization** radio button to allow the entire organization to access this content pack.

5. Enter the desired title for the content pack in the **Title** text box.

6. Enter the desired description for the content pack in the **Description** text box.

7. Click the **Upload** icon to upload an image for the content pack, as shown in Figure 4.84.

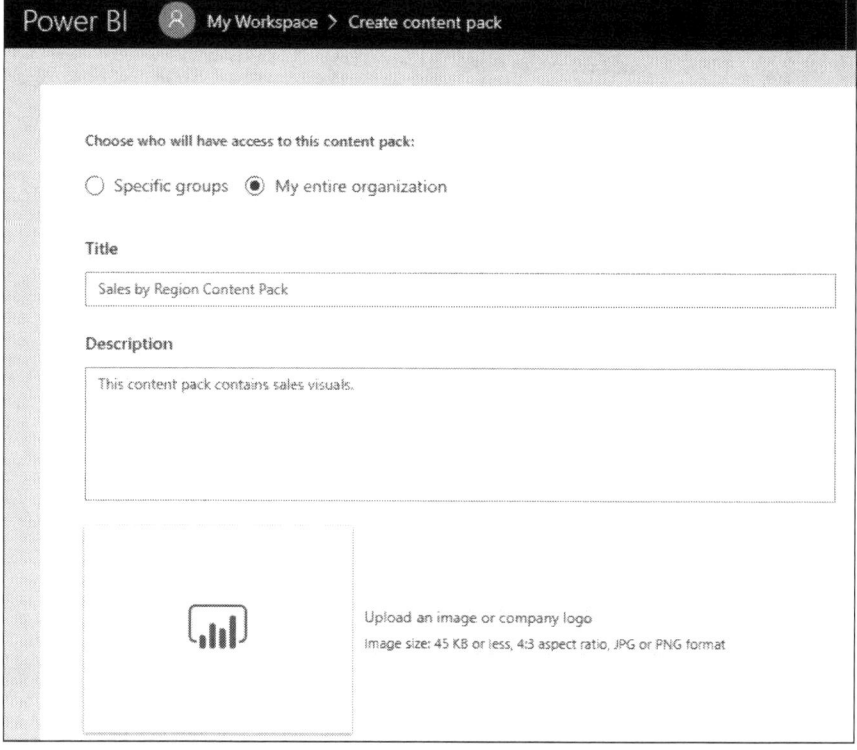

Figure 4.84: Clicking the Upload link

The **Open** dialog box appears.

8. Navigate to the location where the image is located.
9. Select the image.
10. Click the **Open** button to upload the image, as shown in Figure 4.85.

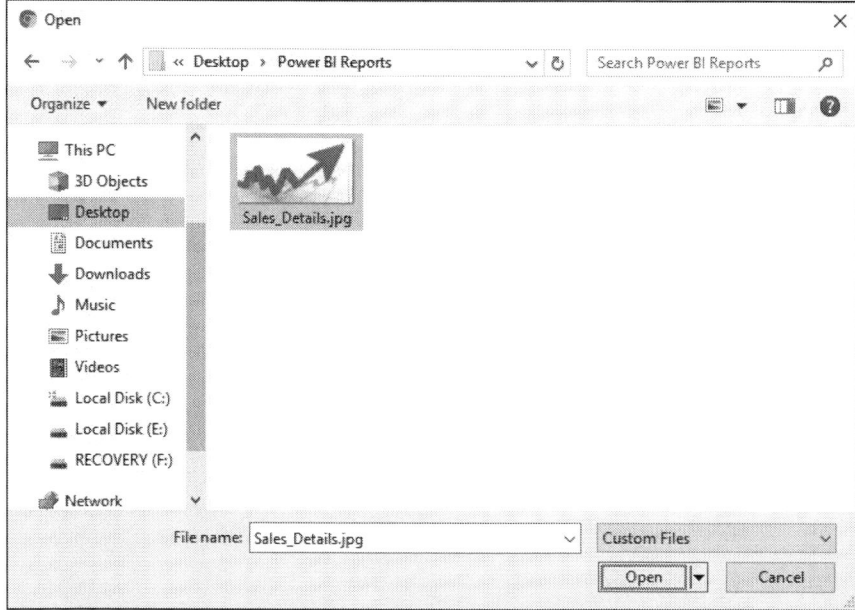

Figure 4.85: Uploading an image

The selected image is uploaded.

11. Select the desired report to be published from the **Reports** list box. The related dataset is selected automatically under the **Datasets** list box.
12. Click the **Publish** button, as shown in Figure 4.86.

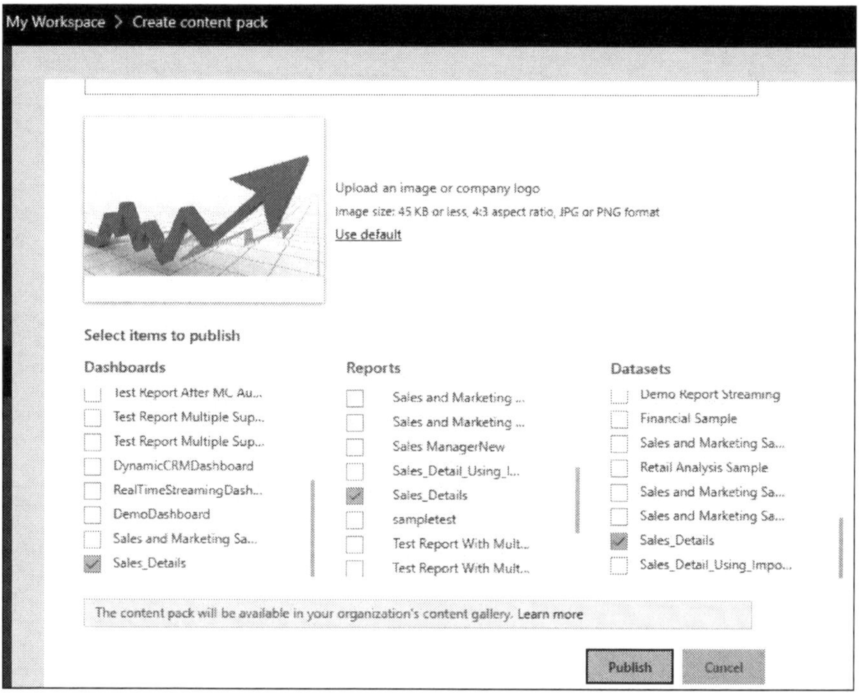

Figure 4.86: Publishing a content pack

A pop-up appears stating that the content pack has been published successfully and added to the organizations content gallery.

You can view the content pack by performing the following steps:
1. Click the **Settings** icon. A drop-down menu appears.
2. Select the **View content pack** option from the drop-down menu, as shown in Figure 4.87.

Manage personal storage
113 MB of 10 GB used
Create content pack
View content pack
Admin portal
Manage gateways
Settings
Manage embed codes

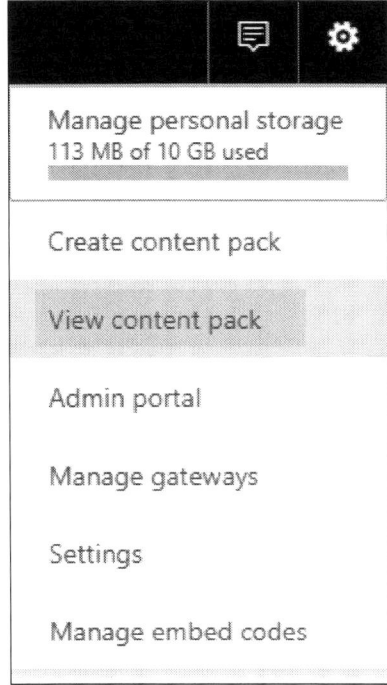

Figure 4.87: Selecting the View content pack option

The **View content pack** page appears listing the content packs, as shown in Figure 4.88.

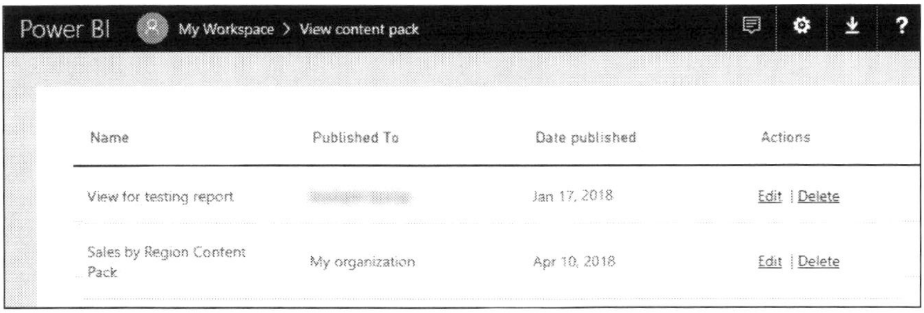

Power BI
My Workspace > View content pack
Name
Published To
Date published
Actions
View for testing report
Jan 17, 2018
Edit | Delete
Sales by Region Content Pack
My organization
Apr 10, 2018
Edit | Delete

Figure 4.88: Viewing content pack

Summary

This chapter outlined the process for integrating PostgreSQL with Power BI. It provided in-depth information about creating a report in Power BI that takes data from the database located in PostgreSQL. We outlined the step-by-step process to take data from the data source, create reports, and then publish the report to Power BI Service. This chapter also reviewed data refresh wherein we discussed gateway setup and configuration of a scheduled refresh. And last, we outlined the process for creating and viewing content packs.

Chapter 5
Power BI on ERP Applications

As discussed earlier, Power BI can be integrated with a number of data sources. One of the most prominent features of Power BI is its ability to integrate with Dynamics CRM, which is a customer relationship management (CRM) solution developed by Microsoft. The ability of Microsoft Dynamics CRM to enhance customer relationships for any organization make it a perfect choice for entrepreneurs as compared to other sources. You can easily integrate Power BI with Microsoft Dynamics CRM through the simple user interface. Users can create simple and intuitive reports in Power BI based on data from Microsoft Dynamics CRM. This chapter provides an in-depth review of Power BI integration with Microsoft Dynamics CRM.

Defining CRM

CRM manages interactions between a company and its customers. It belongs to the data-driven solution category and enhances the interaction and business-driven capabilities of companies with their customers. CRM works as an integrated system for managing customer relationships, tracking sales, and producing data.

Functions of a CRM Solution

The most important functions of a CRM solution are listed below:
-   It simplifies processes and enhances profitability in different divisions including sales, marketing, and service divisions.
-   It provides a multidimensional platform that stores everything related to the development and enhancement of customer relationships.
-   It helps in the growth of opportunities and revenue by building business relationships.
-   It centralizes all customer information.
-   It automates marketing connections.
-   It stipulates business intelligence.
-   It helps in tracking sales opportunities.
-   It empowers responsive customer service.

DOI 10.1515/9781547400720-005

Microsoft Dynamics CRM

To compete with other CRM solutions developed by different vendors, Microsoft introduced its own CRM software and named it Microsoft Dynamics CRM. The defining goal and functionality of Microsoft Dynamics CRM is to improve customer relationships for all types of organizations. It primarily focuses on different divisions/sectors including Sales, Marketing, and Customer Service. Microsoft Dynamics CRM is based on the extended relationship management (xRM) platform, which allows partners to customize it through the .NET based framework. Microsoft Dynamics CRM provides complete support such that CRM apps can be used on mobiles and tablets.

Benefits of Microsoft Dynamics CRM

Although there are several CRM solutions available in the market, Microsoft Dynamics CRM takes precedence over other solutions due to its unique features such as support of deployment models, easy integration with other stacks provided by Microsoft, etc. Some unique features and benefits of Microsoft Dynamics CRM over its competitors are as follows:
- It is easy to use due to its simple user interface (UI).
- It provides different deployment models including on-premises, partner-hosted, hybrid, and online depending upon business structure and needs.
- It supports "point and click" configuration, which eliminates the need to open Microsoft Visual Studio separately to customize the CRM deployment.
- Different licensing options are available per business requirements.
- It supports reporting features that stipulate valuable data insights. It also allows organizations to use SQL Server Reporting Services (SSRS) to create reports.
- It provides the duplicate detection feature and makes the process of importing data from different data sources easy.
- It supports several languages and currencies.
- It is a valuable product that balances both price and features.
- It integrates with commonly used applications including Microsoft Office and the Microsoft Server stack.
- It works on the xRM framework, which makes it a versatile platform for creating custom line-of-business (LOB) applications.

Sign Up for Microsoft Dynamics CRM Online

Microsoft allows organizations to use Microsoft Dynamics CRM solution on a trial basis, and then requires use of paid services as per business requirements.

Perform the following steps to sign up for a trial:
1. Visit the following link:
    https://signup.microsoft.com/Signup?OfferId=bd569279-37f5-4f5c-99d0-425873bb9a4b&dl=DYN365_ENTERPRISE_PLAN1&Culture=en-us&Country=us&flight=AdminOnCustomization&ali=1

The **Dynamics 365** wizard appears with the **Welcome, let's get to know you** dialog box.
2. Select the desired country from the **Country** drop-down list, as shown in Figure 5.1.

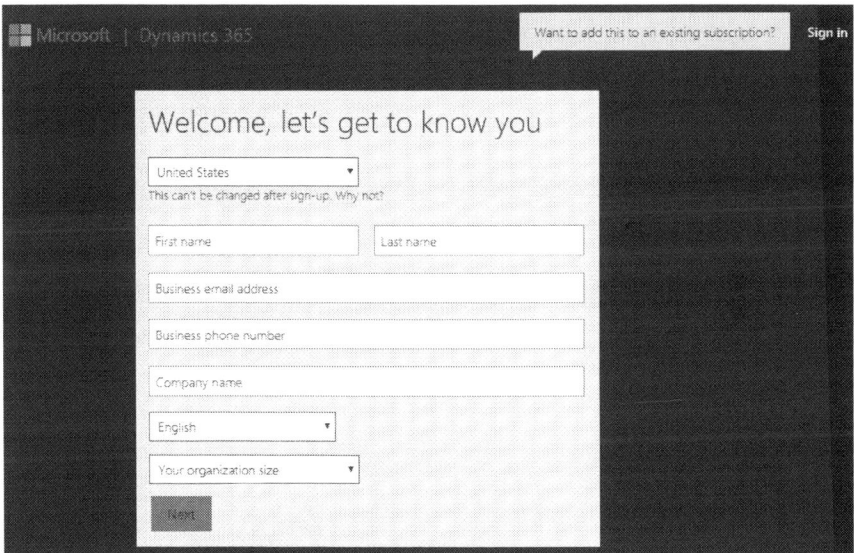

Figure 5.1: Selecting the country

3. Enter your first name in the **First name** text box.
4. Enter your last name in the **Last name** text box.
5. Enter your email address in the **Business email address** text box.
6. Enter your phone number in the **Business phone number** text box.
7. Enter your company name in the **Company name** text box.

8. Select the desired language from the **Language** drop-down list.
9. Select the size of the organization from the **Your organization size** drop-down list.
10. Click the **Next** button.

The **Create your user ID** dialog box appears, as shown in Figure 5.2.

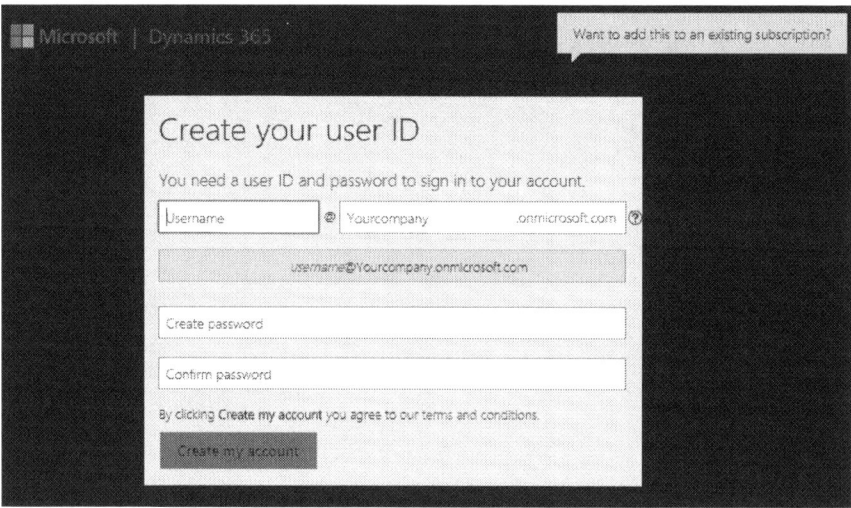

Figure 5.2: The Create your user ID dialog box

11. Enter the desired username in the **Username** text box.
12. Enter the relevant company name in the **Yourcompany** text box.
13. Enter the desired password in the **Create password** text box.
14. Enter the same password in the **Confirm password** text box.
15. Click the **Create my account** button to create the account.

The **Prove. You're. Not. A. Robot.** dialog box appears.
16. Select the **Text me** radio button to get the verification code in text form.
17. Enter a valid phone number in the **Phone number** text box.
18. Click the **Text me** button, as shown in Figure 5.3.

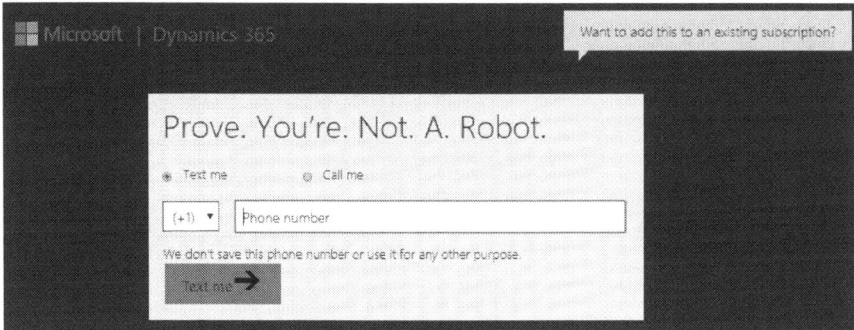

Figure 5.3: Specifying the phone number

The **Prove. You're. Not. A. Robot.** dialog box is redirected.

19. Enter the verification code (as received in the text message sent to the phone number provided above) in the **Enter your verification code** text box.

20. Click the **Next** button, as shown in Figure 5.4.

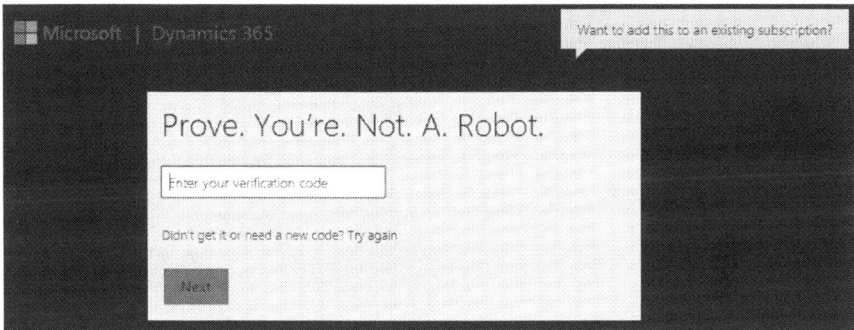

Figure 5.4: Entering verification code

The **Save this info. You'll need it later** dialog box appears.

21. Click the **Set up** button, as shown in Figure 5.5.

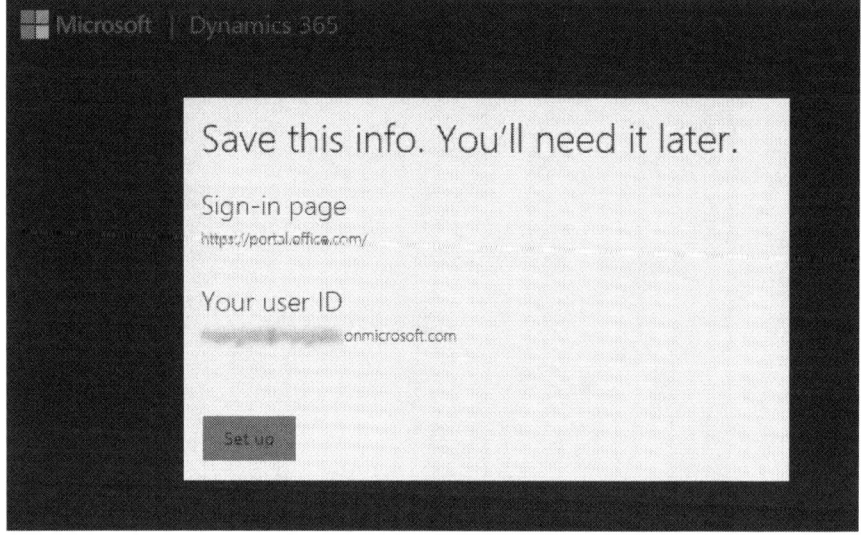

Figure 5.5: Clicking the Set up button

The **Let's get your FREE 30-day trial set up** dialog box appears.

22. Select the **None of these. Don't customize my organization** checkbox.
23. Click the **Complete Setup** button, as shown in Figure 5.6.

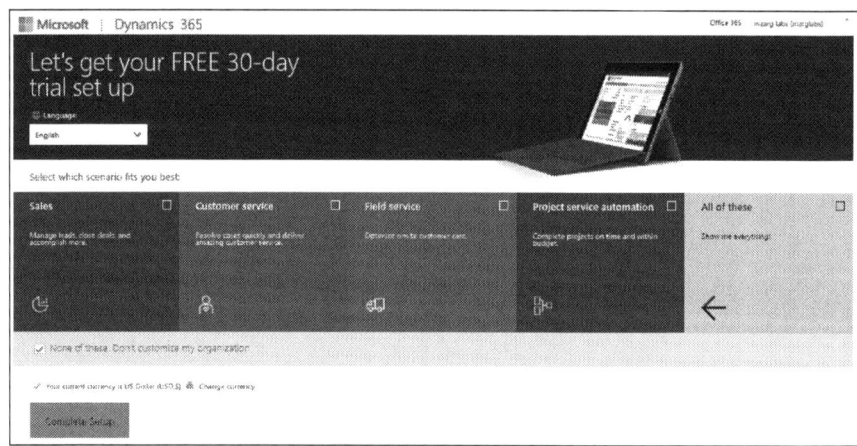

Figure 5.6: The Let's get your FREE 30-day trial set up dialog box

After clicking the **Complete Setup** button, you are redirected to your new CRM Online trial, which has a sample dataset already installed. You can access different sections of CRM by using the navigation options at the top, as shown in Figure 5.7.

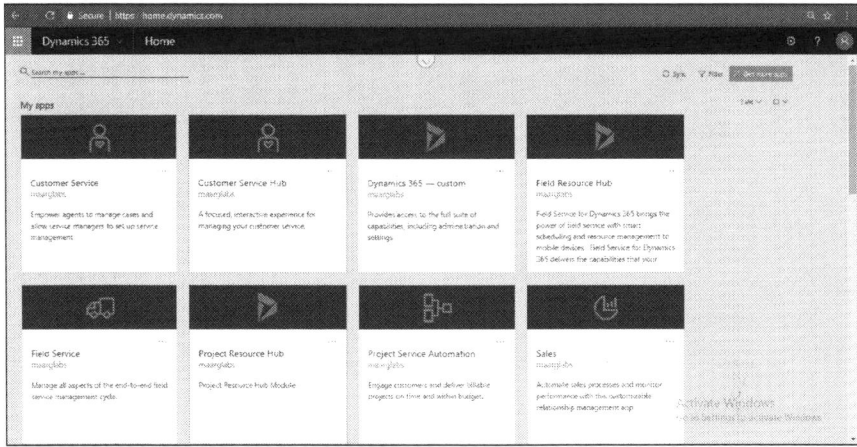

Figure 5.7: Dynamics 365 Home page

Creating Sample Data for Dynamics CRM

In Microsoft Dynamics CRM Online, sample data is already installed. However, in the on-premises version, data is not installed by default. With the on-premises version you need to install the sample data to start working with it.

Perform the following steps to install sample data:
1. Navigate to your CRM tenant.
2. Select the desired category.
3. Click the **Settings** button.
4. Click the **Data Management** option, as shown in Figure 5.8.

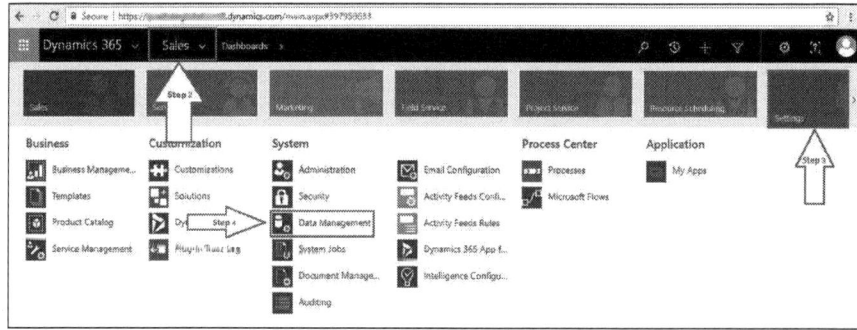

Figure 5.8: Clicking the Data Management option

The **Data Management** page appears.

5. Select the **Sample data** link, as shown in Figure 5.9.

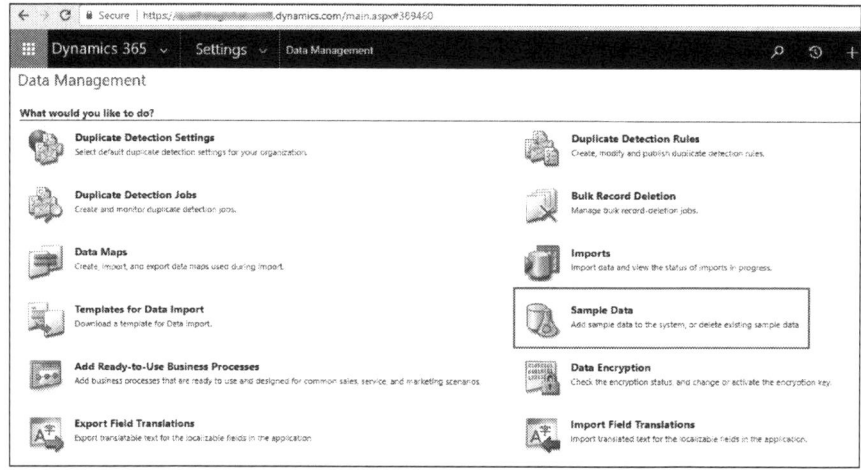

Figure 5.9: Selecting the Sample data link

Sample tables are created in CRM, as shown in Figure 5.10.

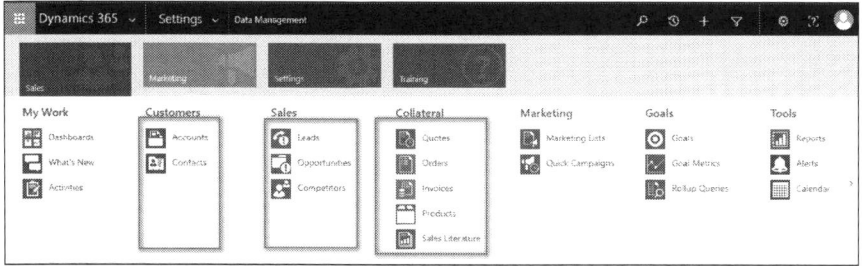

Figure 5.10: Displaying sample data

Getting Data into Power BI

The primary task in the process of creating a report in Power BI is retrieving data from the respective data source, that is, Dynamics CRM in this case.

Perform the following steps to import data into Power BI:

1. Open Power BI desktop.
2. Log in with an organizational account.
3. Click the **Get Data** button under the **External data** section of the **Home** tab, as shown in Figure 5.11.

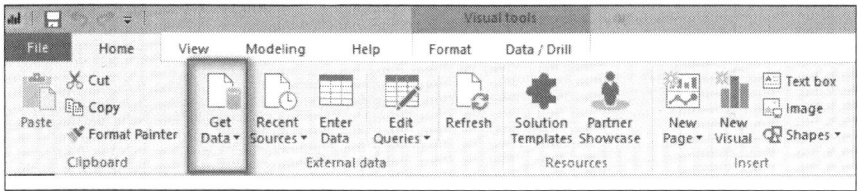

Figure 5.11: Clicking the Get Data button

The **Get Data** dialog box appears.

4. Select the **All** option from the left pane.
5. Select the **Dynamics 365 (online)** option from the right pane.
6. Click the **Connect** button, as shown in Figure 5.12.

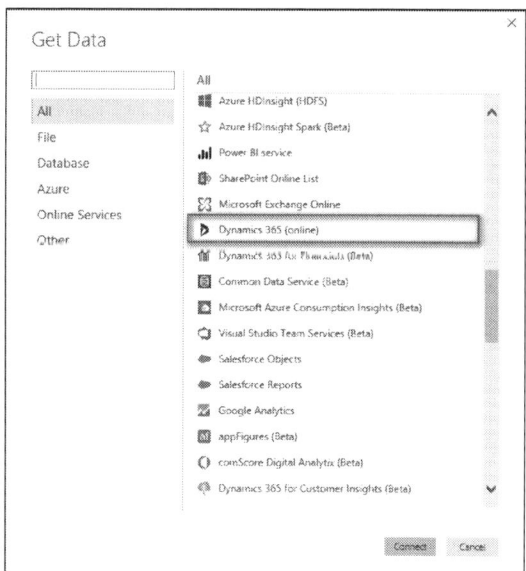

Figure 5.12: The Get Data dialog box

The **Dynamics 365 (online)** dialog box appears.

7. Select the **Basic** radio button.
8. Enter the URL of Dynamics CRM in the **Web API URL** text box.
9. Click the **OK** button to connect with Dynamics CRM, as shown in Figure 5.13.

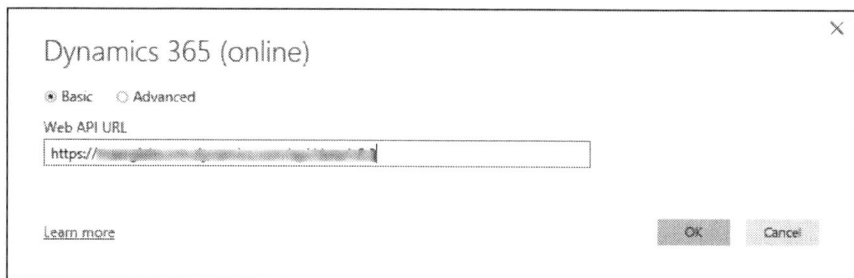

Figure 5.13: The Dynamics 365 (online) dialog box

The **Navigator** window appears, which provides a list of tables that can be loaded to Power BI Desktop.

10. Select the desired tables from the **Navigator** window.
11. Click the **Load** button to load the tables into Power BI Desktop, as shown in Figure 5.14.

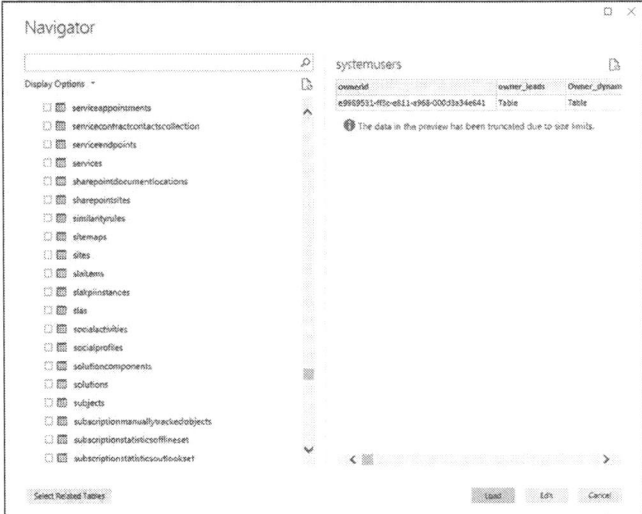

Figure 5.14: The Navigator window

The selected tables are imported into Power BI Desktop and the name of the imported tables display under the FIELDS pane, as shown in Figure 5.15.

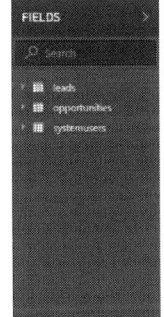

Figure 5.15: The FIELDS pane

Some of the default tables are listed as follows:
- Accounts
- Audits
- Contacts
- Opportunities
- Lead source
- Leads

- System users
- Ratings
- Teams

You may also need to add additional tables to perform specific tasks or create specific reports.

Creating Reports

As discussed earlier, a report is a collection of visuals. In this chapter, we have created a .pbix file that contains the following four reports:
- Leads Revenue by Employee
- Leads Revenue by Company
- Loans by Category
- Loans Summary

Leads Revenue by Employee

The Leads Revenue by Employee report shows the leads revenue generated by employee. We have used different filters and slicers in this report, as shown in Figure 5.16.

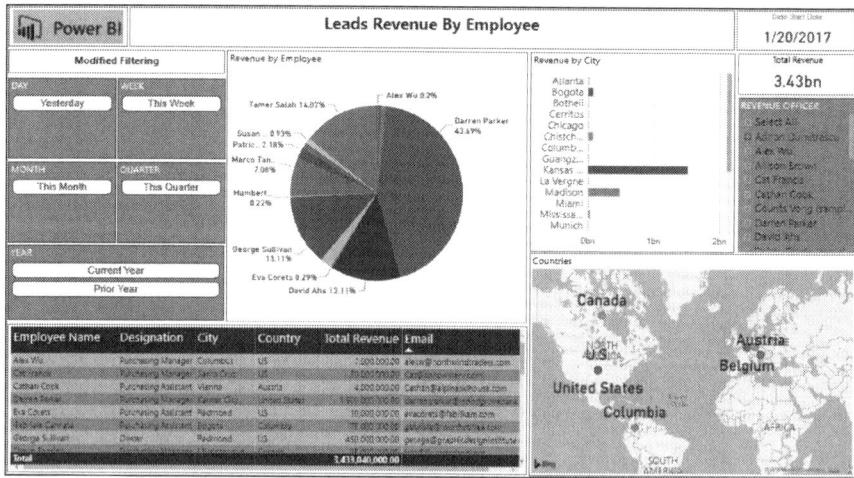

Figure 5.16: The Leads Revenue by Employee report

The primary visuals used in the report above are listed as follows:
- Chiclet slicers
- Pie chart
- Clustered bar chart
- Slicer
- Table
- Map

Chiclet Slicers

Similar to other slicers, a chiclet slicer is used to apply filters on data. In the above report, we have created five chiclet slicers, as shown in Figure 5.17.

Figure 5.17: Chiclet slicers

The mapping details of the DAY chiclet slicer are as follows:

```
DAY =

if(FORMAT(leads[createdon], "MM-DD-YYYY") = FORMAT(TODAY(), "MM-DD-YYYY"),
"Today",

if(FORMAT(leads[createdon], "MM-DD-YYYY") = FORMAT(TODAY()+1, "MM-DD-YYYY"),
"Tomorrow",
```

```
if(FORMAT(leads[createdon], "MM-dd-YYYY") = FORMAT(TODAY()-1, "MM-dd-YYYY"),
"Yesterday", BLANK())))
```

The mapping details of the WEEK chiclet slicer are as follows:

```
WEEK =

if(leads[WeekNo]=WEEKNUM(TODAY(),21) && YEAR(leads[createdon])=YEAR(TODAY()),"This
Week",

if((WEEKNUM(TODAY(),21)-1) = 0 && leads[WeekNo]=53 && YEAR(leads[createdon])=YEAR(
TODAY())-1,"Last Week",

if(leads[WeekNo]=WEEKNUM(TODAY(),21)-1 &&
YEAR(leads[createdon])=YEAR(TODAY()),"Last Week",

if(leads[WeekNo]=WEEKNUM(TODAY(),21)+1 &&
YEAR(leads[createdon])=YEAR(TODAY()),"Next Week",

if((WEEKNUM(TODAY())+1) = 54 && leads[WeekNo]=1 &&
YEAR(leads[createdon])=YEAR(TODAY())+1,"Next Week",

BLANK()))))))
```

The mapping details of the MONTH chiclet slicer are as follows:

```
Month Duration =

if(FORMAT(leads[createdon], "MM-YYYY") = FORMAT(TODAY(), "MM-YYYY"), "This Month",

if(FORMAT(leads[createdon], "MM-YYYY") = FORMAT(EOMONTH(TODAY(), -1), "MM-YYYY"),
"Last Month",

if(FORMAT(leads[createdon], "MM-YYYY") = FORMAT(EOMONTH(TODAY(), 1), "MM-YYYY"),
"Next Month", BLANK())))
```

The mapping details of the QUARTER chiclet slicer are as follows:

```
QUARTER = if(leads[QuarterNo]=ROUNDUP(MONTH(TODAY())/3,0) &&
YEAR(TODAY())=leads[Year],"This Quarter",

if(leads[QuarterNo]=(ROUNDUP(MONTH(TODAY())/3,0)-1) &&
YEAR(TODAY())=leads[Year],"Last Quarter",

if(leads[QuarterNo]=4 && (ROUNDUP(MONTH(TODAY())/3,0)-1)=0 && YEAR(TODAY())-
1=leads[Year],"Last Quarter",

if(leads[QuarterNo]=(ROUNDUP(MONTH(TODAY())/3,0)+1) &&
YEAR(TODAY())=leads[Year],"Next Quarter",
```

```
if(leads[QuarterNo]=1 && (ROUNDUP(MONTH(TODAY())/3,0)+1)=5 &&
YEAR(TODAY())+1=leads[Year],"Next Quarter",

BLANK())))))
```

The mapping details of the YEAR chiclet slicer are as follows:

```
YearDuration =

if(YEAR(leads[createdon]) = YEAR(TODAY()), "Current Year",

if(YEAR(leads[createdon]) = CALCULATE(YEAR(TODAY())-1), "Prior Year",

if(YEAR(leads[createdon]) = CALCULATE(YEAR(TODAY())+1),"Next Year", BLANK())))
```

Pie Chart Visual

A pie chart visual is a circle divided into slices and displays data in numerical proportions. This pie chart visual shows the Revenue by Employee data, as shown in Figure 5.18.

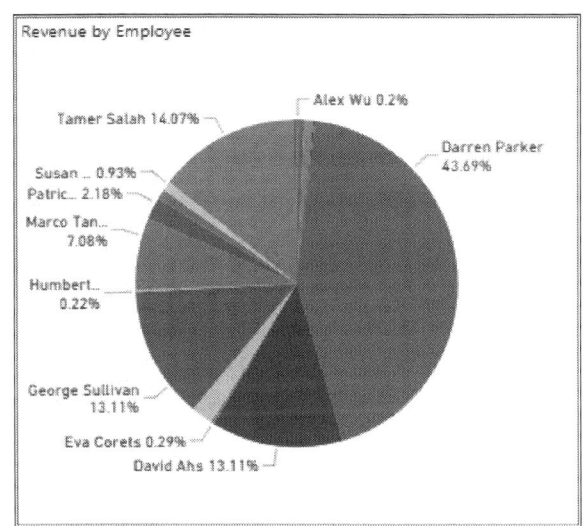

Figure 5.18: The Revenue by Employee pie chart

Clustered Bar Chart

A clustered bar chart is a simple bar chart where different graph bars are placed next to each other. The REVENUE BY CITY chart displays revenue data filtered by city, as shown in Figure 5.19.

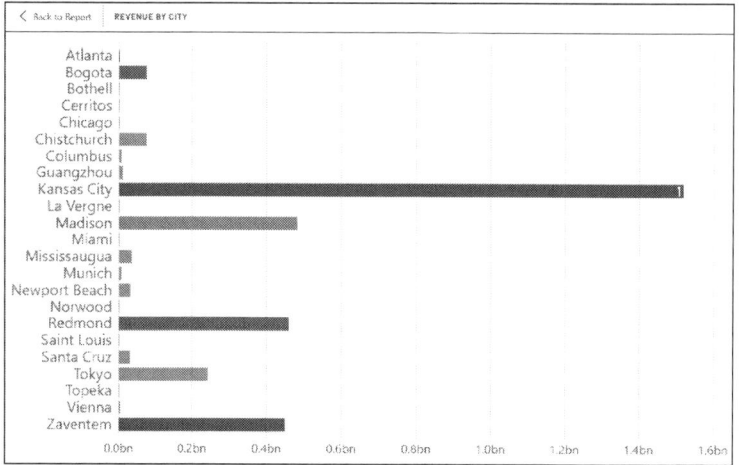

Figure 5.19: The REVENUE BY CITY clustered bar chart

Slicer

A Slicer visual acts as a filter that is applied on other report visuals to provide filtered results. The REVENUE OFFICER slicer lists the name of revenue officers, and filters other visuals based on the selected officer in the "Leads Revenue by Employee" report, as shown in Figure 5.20.

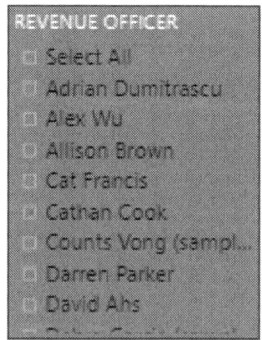

Figure 5.20: The REVENUE OFFICER Slicer visual

Map

The Map visual highlights specific or important locations on a map. In this report, this visual highlights the countries, as shown in Figure 5.21.

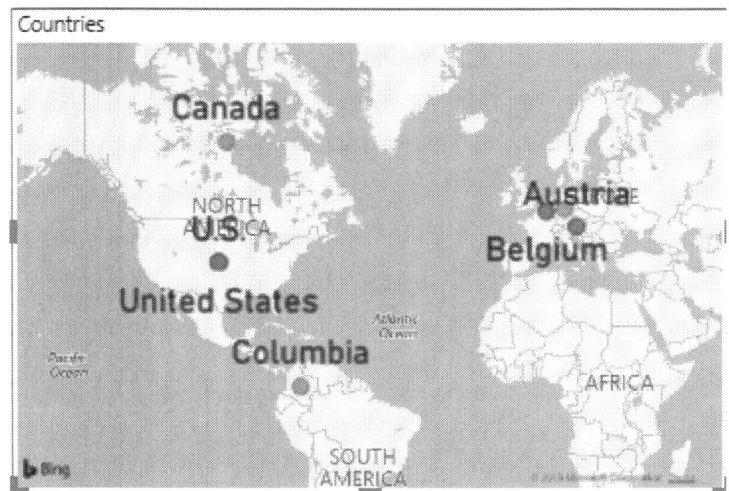

Figure 5.21: The Map visual

Table

The Table visual contains rows and columns. In this report, the Table visual displays information including Employee Name, Designation, City, Country, Total Revenue, and Email, as shown in Figure 5.22.

Employee Name	Designation	City	Country	Total Revenue	Email
Alex Wu	Purchasing Manager	Columbus	US	7,000,000.00	alexw@northwindtraders.com
Cat Francis	Purchasing Manager	Santa Cruz	US	30,000,000.00	Cat@cohowinery.com
Cathan Cook	Purchasing Assistant	Vienna	Austria	4,000,000.00	Cathan@alpineskihouse.com
Darren Parker	Purchasing Manager	Kansas City	United States	1,500,000,000.00	darren.parker@woodgrovebank.
Eva Corets	Purchasing Assistant	Redmond	US	10,000,000.00	evacorets@fabrikam.com
Gabriele Cannata	Purchasing Assistant	Bogota	Columbia	75,000,000.00	gabriele@fourthcoffee.com
George Sullivan	Owner	Redmond	US	450,000,000.00	george@graphicdesigninstitute.
	Purchasing Manager	Mississauga	Canada	25,000,000.00	
Total				**3,433,040,000.00**	

Figure 5.22: The Table visual

Leads Revenue by Company

The "Leads Revenue by Company" report displays the leads revenue filtered through companies. This report contains chiclet slicers, a stacked area chart, a pie chart, the slicer filter, and a table, as shown in Figure 5.23.

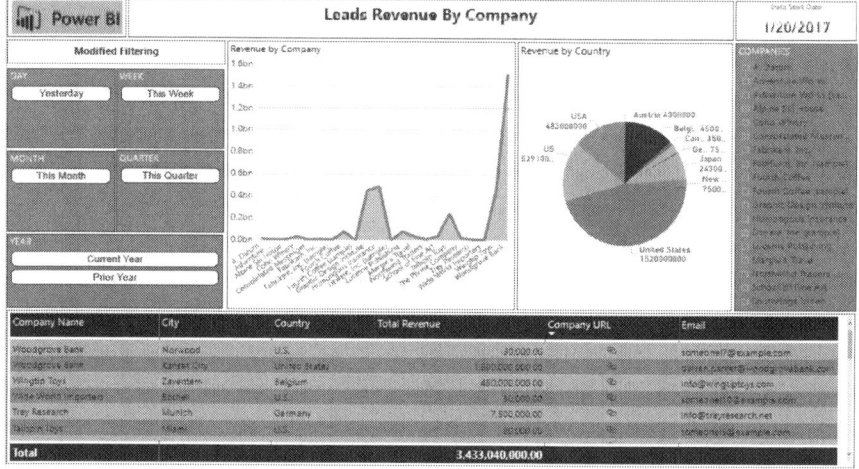

Figure 5.23: The Leads Revenue by Company report

The Leads Revenue by Company report contains the following visuals:
– Chiclet slicers
– Stacked area chart
– Pie chart
– Slicer visual
– Table

Stacked Area Chart

A stacked area chart is similar to an area chart that displays the values of numerous groups on the same visual. The "Leads Revenue by Company" report contains the "Revenue by Company" stacked area chart, as shown in Figure 5.24.

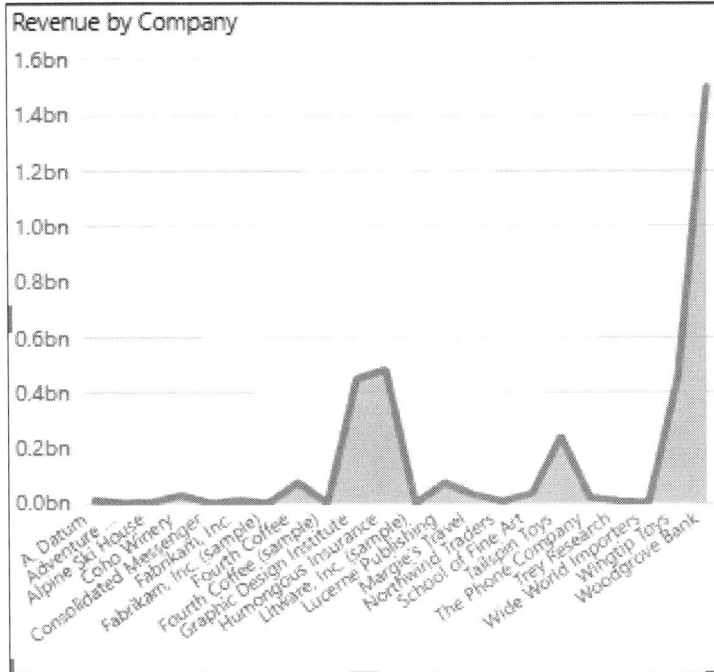

Figure 5.24: Stacked area chart

Slicer

A Slicer visual acts as a filter that is applied on other report visuals to provide filtered results. The COMPANIES slicer lists the name of companies and thus filters the revenue status of each company in the "Leads Revenue by Company" report, as shown in Figure 5.25.

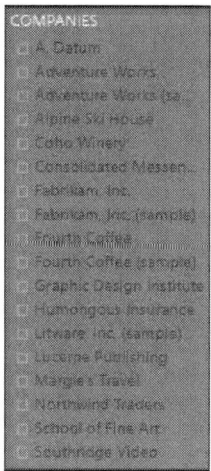

Figure 5.25: The COMPANIES slicer

Loans by Category

The "Loans by Category" report displays data related to loans filtered by product category. This report also contains filters and slicers, as shown in Figure 5.26.

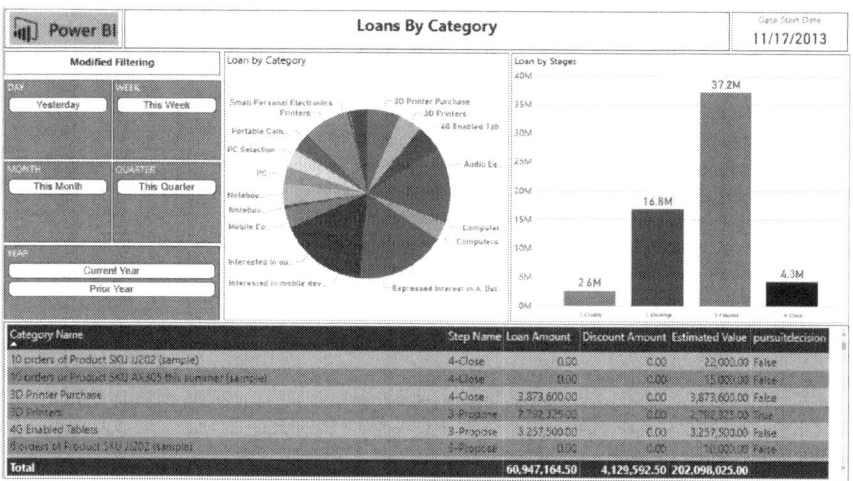

Figure 5.26: The Loans by Category report

The "Loans by Category" report contains the following visuals:
- Chiclet slicers
- Pie chart
- Clustered column chart
- Table

Clustered Column Chart

A clustered column chart is a column chart that compares values across defined categories through vertical bars. The "Loan by Stages" clustered column chart displays loan values per Step Name category, as shown in Figure 5.27.

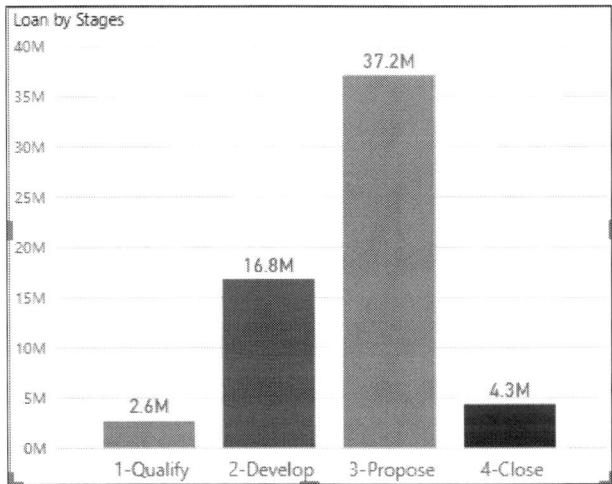

Figure 5.27: Clustered column chart

Loans Summary

The "Loans Summary" report displays a summary of total loans. This report contains chiclet slicers, a donut chart, a card grid, PBI_CV, and a table visual, as shown in Figure 5.28.

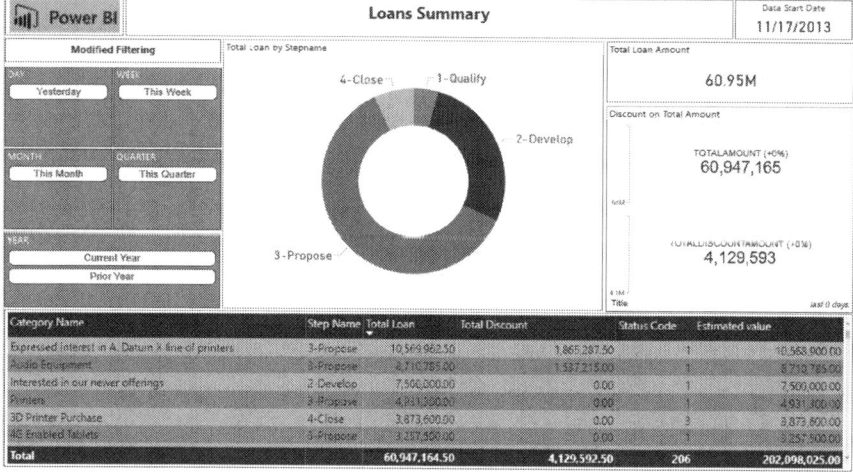

Figure 5.28: The Loans Summary report

Donut Chart

Donut (or doughnut) chart is similar to a pie chart that contains slices wherein each section depicts a value. The "Total Loan by Stepname" donut chart is shown in Figure 5.29.

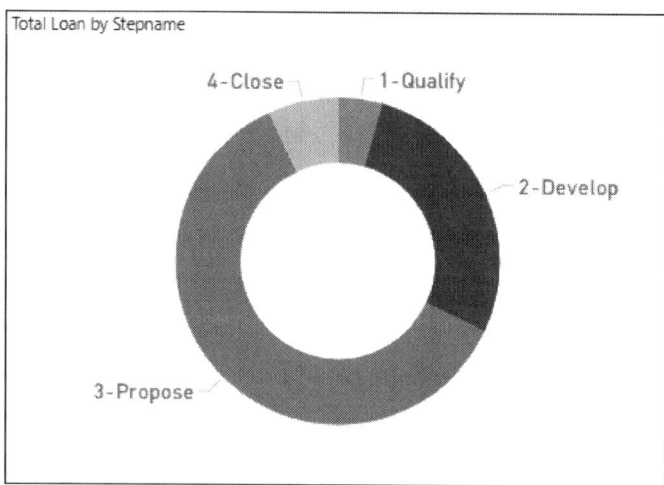

Figure 5.29: The "Total Loan by Stepname" donut chart

Deep Linking in Power BI

Deep linking allows users to view information available from the Dynamics CRM portal by clicking the available link. This link redirects the user to the linked URL. Figure 5.30 shows the deep linking feature applied on the values under the Company URL column.

Company Name	City	Country	Total Revenue	Company URL	Email
Woodgrove Bank	Norwood	U.S.	30,000.00		someone17@example.com
Woodgrove Bank	Kansas City	United States	3,400,000,000.00		daren.parents@woodgrovebank.com
Wingtip Toys	Zaventem	Belgium	450,000,000.00		info@wingtiptoys.com
Wide World Importers	Bothell	U.S.	30,000.00		someone18@example.com
Trey Research	Munich	Germany	7,500,000.00		info@treyresearch.net
Tailspin Toys	Miami	U.S.	80,000.00		someone15@example.com
Total			**3,433,040,000.00**		

Figure 5.30: Using the deep linking feature

Adding a New User

You can add a new user in Microsoft Dynamics CRM Online trial. Perform the following steps to add a new user:

1. Navigate to the Dynamics 365 admin portal.
2. Click the **Add a user** link under the **Active users** section, as shown in Figure 5.31.

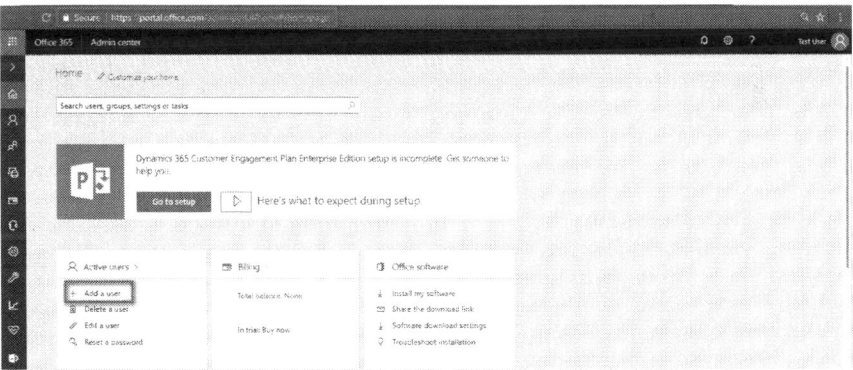

Figure 5.31: Adding a new user

The **New user** window appears, as shown in Figure 5.32.

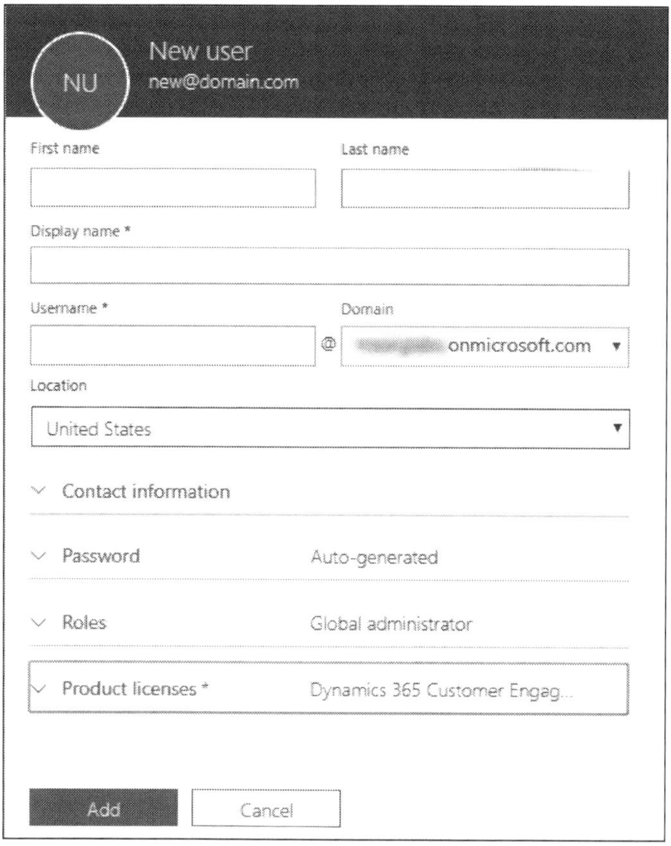

Figure 5.32: The New user window

3. Enter the desired first name in the **First name** text box.
4. Enter the desired last name in the **Last name** text box.
5. Enter the display name in the **Display name** text box.
6. Enter the desired user name in the **Username** text box.
7. Select the desired location from the **Location** drop-down list.
8. Expand the **Roles** section. Different options related to the **Roles** section appear.
9. Select the desired radio button under the **Roles** section to assign a role to the user, as shown in Figure 5.33.

Figure 5.33: Assigning a role

10. Expand the **Product licenses** section.
11. Assign the desired licenses to the new user.
12. Click the **Add** button, as shown in Figure 5.34.

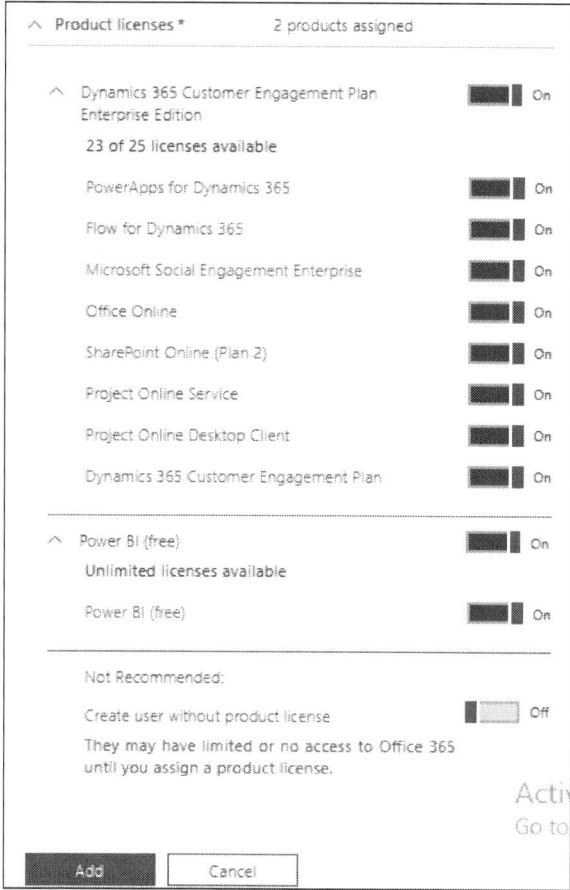

Figure 5.34: Adding a new user

A confirmation window appears with the status "User was added," as shown in Figure 5.35.

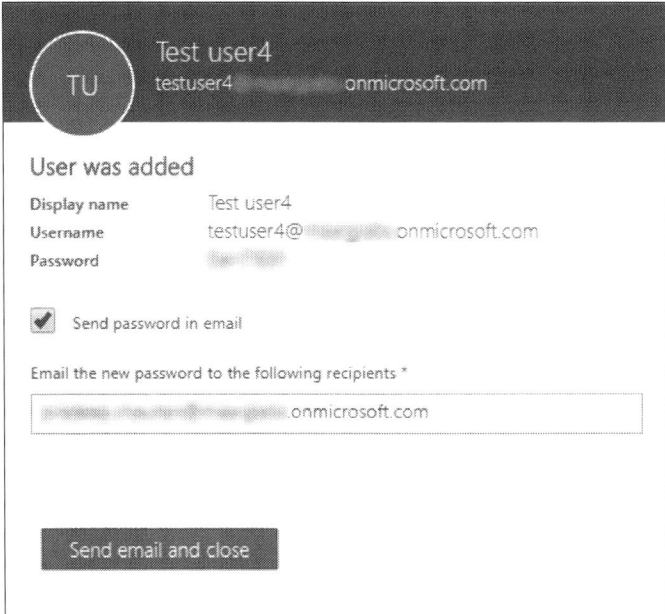

Figure 5.35: Displaying a confirmation window

Row-Level Security in Power BI

Row-level security (RLS) is a feature of Power BI that filters content based on a user's role. Simply speaking, RLS is used to impose data access restrictions for available users. You can apply filters on data at the row level. These filters can be specified under available roles.

Several considerations related to RLS are as follows:
- RLS can be organized for data models that are brought into Power BI through Power BI Desktop.
- RLS can be defined on datasets that are created through the DirectQuery option in data sources including SQL Server.

You can create roles and rules in Power BI Desktop. These roles definitions are automatically published to Power BI Service when you publish your report from Power BI Desktop to Power BI Service. You can apply filters to these roles through DAX expressions.

Defining Roles and Rules

As discussed earlier, the roles and rules defined in Power BI Desktop are automatically published to Power BI Service when you publish your report to Power BI Service.

Perform the following steps to define security roles in Power BI Desktop:
1. Import data into Power BI Desktop using the **Get data** option.
2. Select the **Modeling** tab.
3. Click the **Manage Roles** button under the **Security** section, as shown in Figure 5.36.

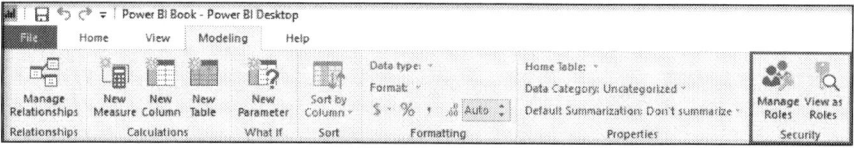

Figure 5.36: Clicking the Manage Roles button

The **Manage roles** dialog box appears.
4. Click the **Create** button under the **Roles** section to create a new role, as shown in Figure 5.37.

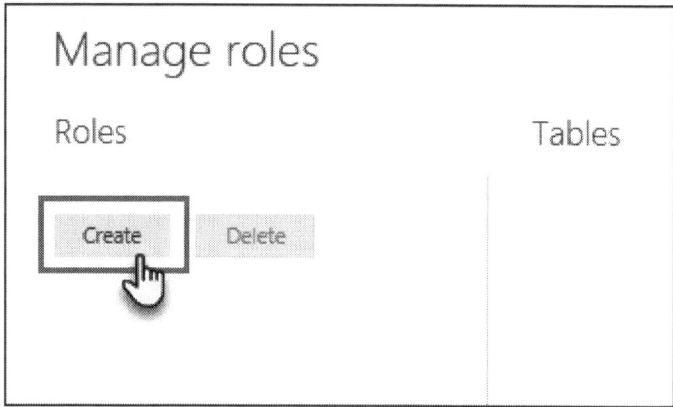

Figure 5.37: The Manage roles dialog box

A text box appears with the text "New role."

5. Replace the text with the desired name of the role. In our case, we have replaced the text with "Manager." Similarly, we have created one more role with the name "Officer."
6. Select the desired table from the **Tables** section for which you want to apply a DAX rule.
7. Enter the desired DAX rule for the selected table in the **Table filter DAX expression** text area. In our case, we have three tables, namely "leads," "opportunities," and "systemusers." From these tables, we have used the "leads" and "opportunities" tables. We have written the following DAX rules for these tables:

For "leads" table:

```
[address1_country]=LOOKUPVALUE(systemusers[address1_
country],systemusers[internalemailaddress],UserName())
```

For "opportunities" table:

```
[Country]=LOOKUPVALUE(systemusers[address1_
country],systemusers[internalemailaddress],UserName())
```

Note
Data Analysis Expressions (DAX) is an expression (a set of functions, constants, and operators) written to apply calculations on data available in your model. The DAX expression should either return a True or False.

8. Click the **OK** icon (✓) next to the **Table filter DAX expression** option to validate the expression, as shown in Figure 5.38.

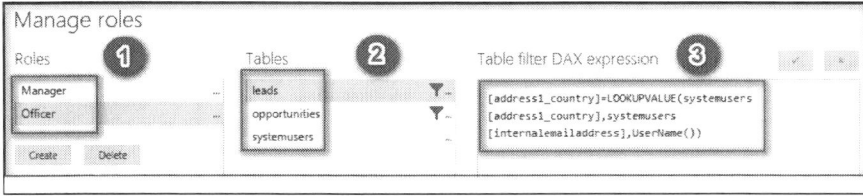

Figure 5.38: Applying and validating DAX rule

9. Click the **Save** button to save the created roles.

Note

In the default configuration, RLS filtering uses single-directional filters irrespective of the direction set for the relationships. However, you can enable bi-directional filtering by selecting the bi-directional option and selecting the Apply security filter in both directions checkbox. You should select the Apply security filter in both directions checkbox at the time of implementing dynamic RLS. In dynamic RLS, RLS is implemented based on user name or login ID.

Validating the Role within Power BI Desktop

After creating the role in Power BI Desktop, the results of the created role can be validated within Power BI Desktop.

Perform the following steps to validate the role within Power BI Desktop:

1. Click the **View as Roles** button under the **Security** group of the **Modeling** tab, as shown in Figure 5.39.

Figure 5.39: Clicking the View as Roles button

The **View as roles** dialog box opens.

2. Select the role to be used as a filter. In our case, we have selected **Officer**.
3. Click the **OK** button, as shown in Figure 5.40.

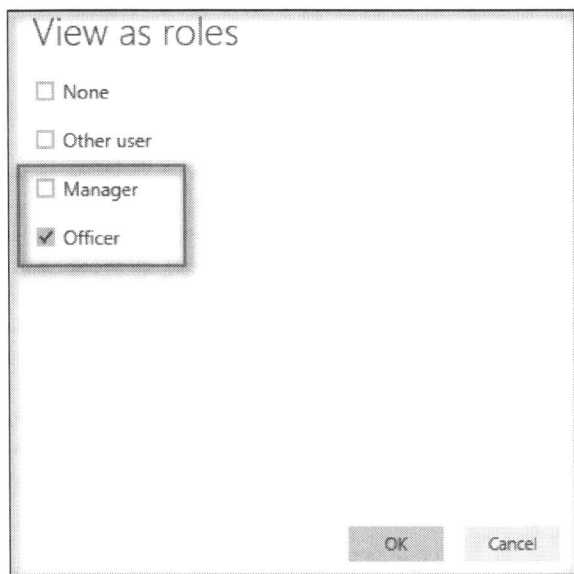

Figure 5.40: The View as roles dialog box

After clicking the **OK** button, the visuals in the report are filtered as per the selected role, as shown in Figure 5.41.

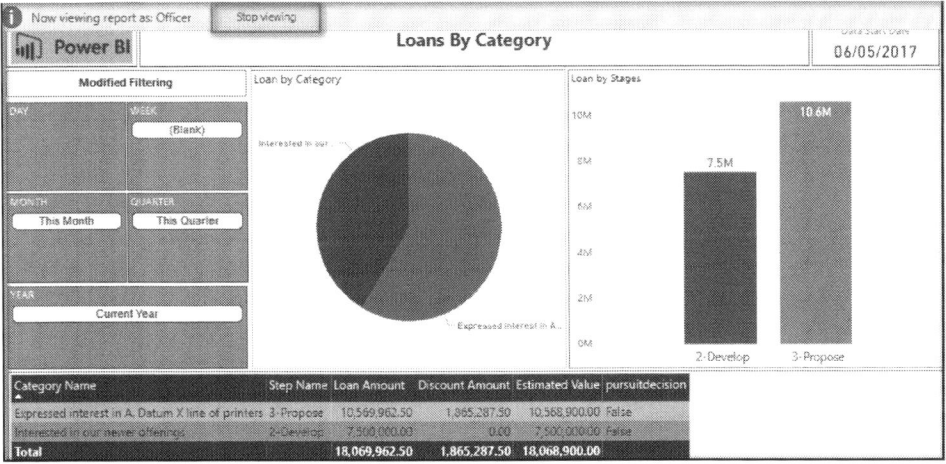

Figure 5.41: Validating the role

Deploying the Report

Once the required roles are created in Power BI Desktop, you can deploy the report containing roles to Power BI Service.

Perform the following steps to deploy the report to Power BI Service:
1. Click the **Publish** button under the **Share** section of the **Home** tab, as shown in Figure 5.42.

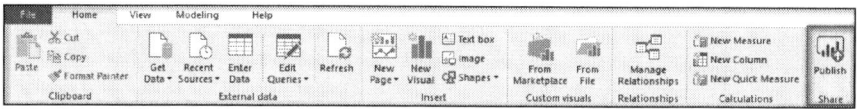

Figure 5.42: Clicking the Publish button

The selected report is deployed to Power BI Service successfully, as shown in Figure 5.43.

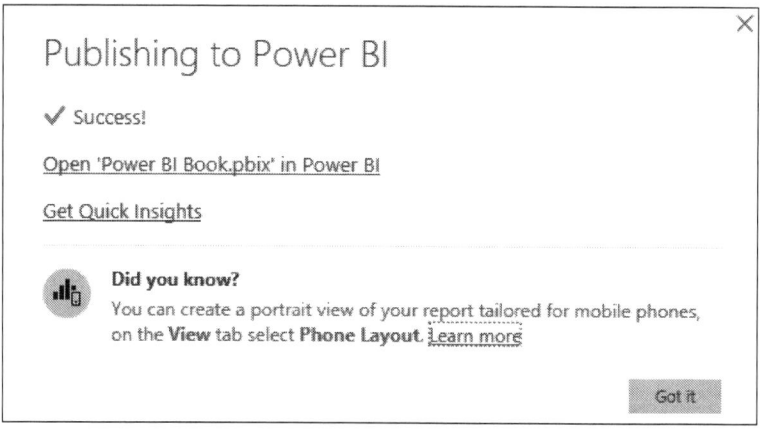

Figure 5.43: Deploying the report

Managing Security

After creating a role in Power BI Desktop and publishing the report to Power BI Service, you can manage RLS on your data model or dataset.

Perform the following steps to manage RLS:

1. Navigate to Power BI Service.
2. Expand the **My Workspace** section.
3. Select the **Ellipsis** icon (...) next to the name of the applicable dataset under the **DATASETS** section. A menu appears.
4. Select the **SECURITY** option from the menu, as shown in Figure 5.44.

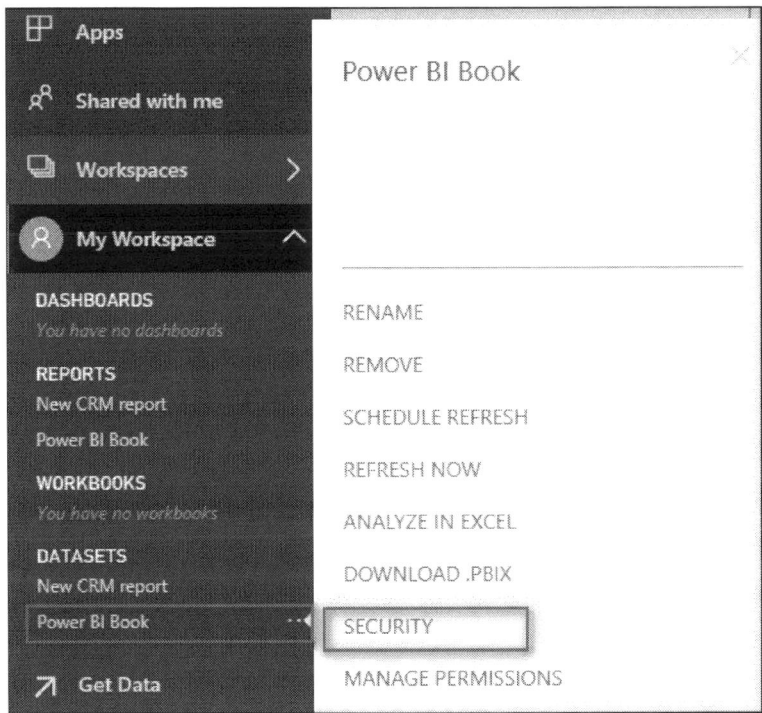

Figure 5.44: Selecting the SECURITY option

Selecting the SECURITY option opens the RLS page wherein the left pane displays the name of the roles created in Power BI Desktop, while the right pane allows you to add members to a role selected in the left pane, as shown in Figure 5.45.

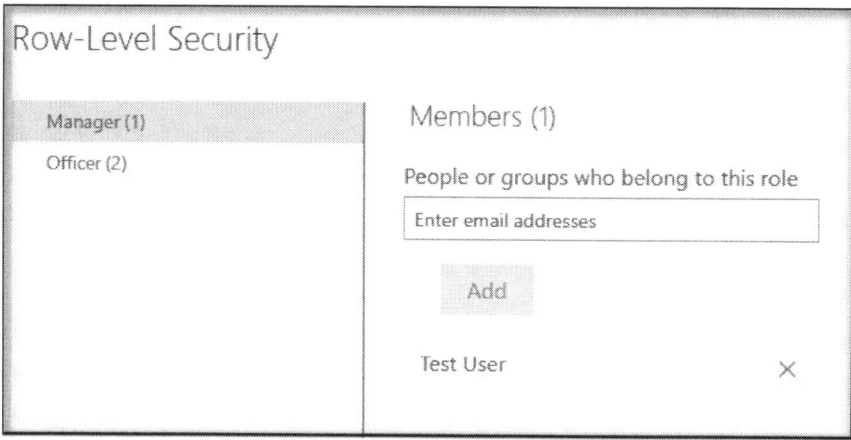

Figure 5.45: The Row-Level Security page

Note

The Security option is available for dataset owners only. In a case where the dataset relates to a group, the Security option is available for group administrators.

Working with Members

After creating a role in Power BI, you need to add members to the specified role to allow them to see the filtered results specified in RLS. This section covers the following topics:
- Adding a member
- Removing a member

Note

Roles are created or modified in Power BI Desktop, while members within the roles are added in Power BI Service.

Adding a Member
Perform the following steps to add a member:
1. Open Power BI Service.
2. Navigate to the dataset that you have published by applying RLS.
3. Click the **Ellipsis** icon (...) next to the name of the dataset. A menu appears.

4. Select the **Security** option from the menu. The **Row-Level Security** page appears.
5. Select the desired role to which you want to add the member.
6. Enter the email address or name of the user that you want to add to the role.

Note
The member that you are adding to the role should belong to your organization. Groups that are created within Power BI can not be added as a member.

---

7. Click the **Add** button, as shown in Figure 5.46.

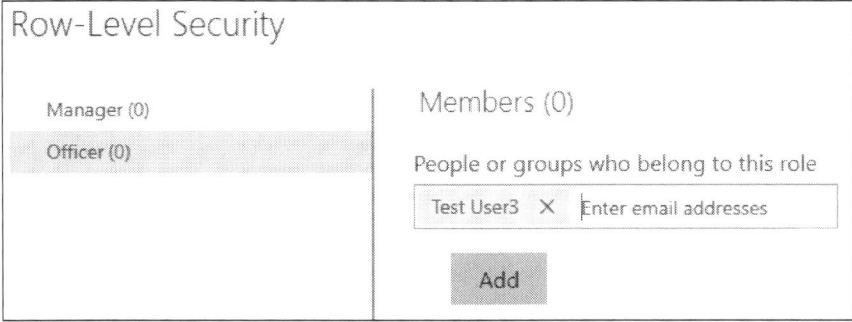

Row-Level Security

Manager (0)	**Members (0)**
Officer (0)	People or groups who belong to this role
	Test User3 ✕ Enter email addresses
	Add

Figure 5.46: Adding a member

8. Click the **Save** button to save the configuration.

Note
The value specified in the parentheses beside the role name represents the number of members belonging to the role.

Removing a Member

Perform the following steps to remove a member:

1. Navigate to the **Row-Level Security** page.
2. Select the desired role from which you want to remove a member.
3. Click the **Close** icon **(X)** next to the name of the member to remove the selected member, as shown in Figure 5.47.

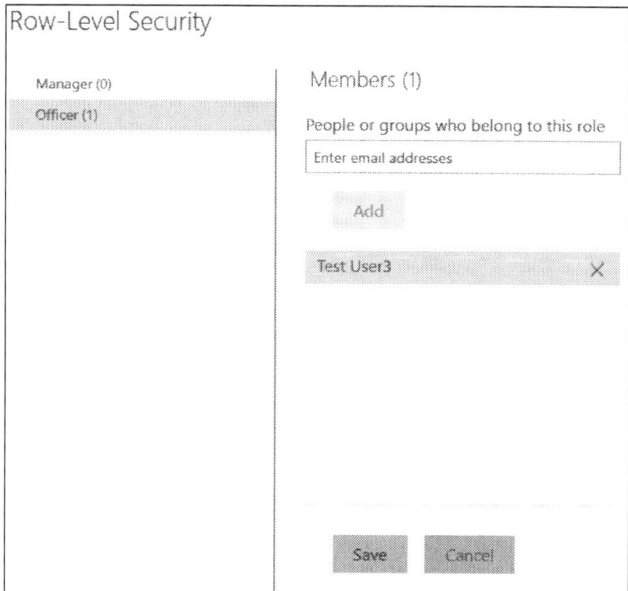

Figure 5.47: Removing a member

4.   Click the **Save** button to save the configuration.

Validating the Role within Power BI Service

Similar to the context of validating the role in Power BI Desktop, you can validate the working of the defined role in Power BI Service.

Perform the following steps to validate the roles within Power BI Service:

1.   Expand the **My Workspace** option.
2.   Click the **Ellipsis** icon **(...)** next to the report name you want to validate. A menu appears.
3.   Select the **Test data as role** option. The report is filtered based on the selected role, as shown in Figure 5.48.

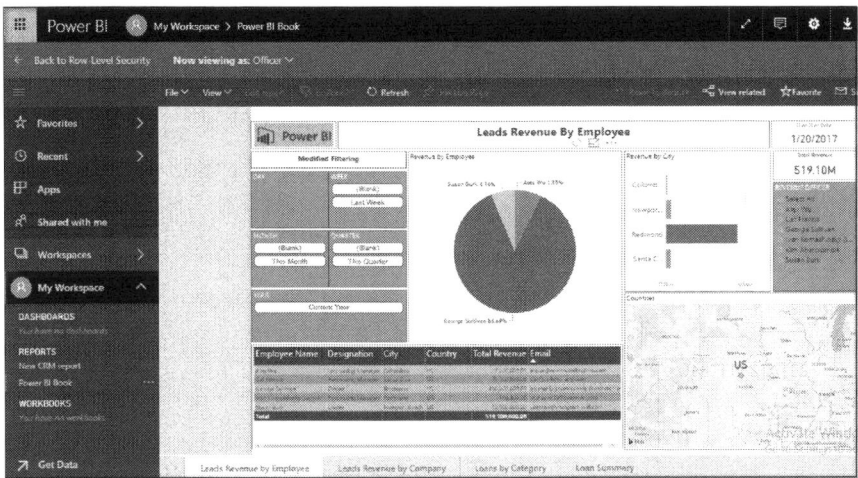

Figure 5.48: Validating the role

You can also filter the report based on other roles by selecting the other role from the **Now viewing as** drop-down list.

Sharing the Report

Once you are done with all your work on the report, you can easily share it with your colleagues.

Perform the following steps to share the report:
1. Open the report you want to share in Power BI Service.
2. Click the **Share** button to share the report, as shown in Figure 5.49.

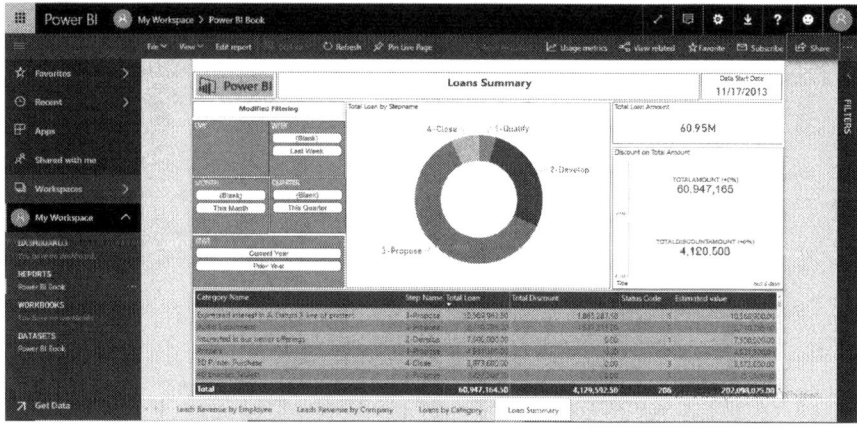

Figure 5.49: Clicking the Share button

The **Share report** dialog box appears.

3. Select the **Share** tab.
4. Enter the email address of the recipient with which you want to share the report in the **Grant access to** text box.
5. Enter a message in the **Include an optional message** text area.
6. Select the **Allow recipients to share your report** checkbox.
7. Select the **Send email notification to recipients** checkbox.
8. Click the **Share** button to share the report link, as shown in Figure 5.50.

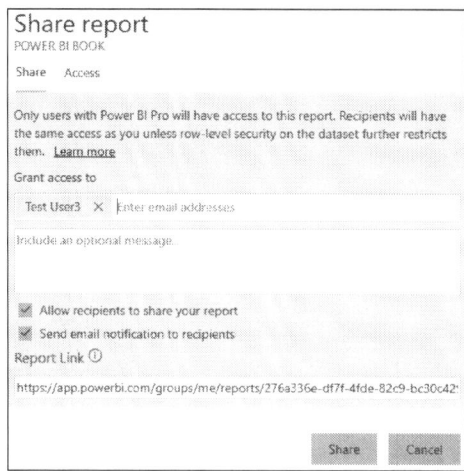

Figure 5.50: Sharing a report

The specified recipient will receive an email message that contains a link to the report, and can open this report by clicking the received link.

Summary

Microsoft Dynamics CRM is a customer relationship management solution based on the xRM platform. There are several benefits of Microsoft Dynamics CRM as compared to other CRM solutions. Some of its key benefits are that it has a very simple and easy to use UI; it has multiple licensing options; it supports reporting features and several languages and currencies. This chapter first outlined the process to sign up for a Microsoft Dynamics CRM Online. Next, it described the complete process of taking data from Dynamics CRM and importing it into Power BI. We created four reports wherein each report contained multiple visuals. New chiclet slicers were also defined in this chapter. Next, we described the deep linking feature that allows users to view the information available on the Dynamics CRM portal by clicking the available link. And last, we summarized the RLS feature within Power BI, which filters content based on a user's role.

Chapter 6
Conclusion

Power BI is a business intelligence and reporting tool that allows users to create intuitive reports. A growing number of organizations are using Power BI as their business analytics solution. According to Gartner's report, by 2020, the companies that are investing in analytics will see their value enhanced compared to those that are not. This chapter provides a summary of all the chapters covered in this book.

Introducing Data Visualization

Data visualization is the concept of presenting data through visuals, such as infographics, charts, Sparkline, and geographic maps, etc. Data visualization provides an efficient and effective way to communicate concepts in general because the human brain processes visualized information more easily than textual information. A few of the most commonly used data visualization tools are Microsoft Power BI, Tableau, and Qlik.

Introduced by Microsoft, Power BI is a business analytics reporting tool used to create interactive business reports. It incorporates several analytics features to provide business insights across an entire organization.

The top two competitors in the list of the business intelligence (BI) and data visualization tools are Power BI and Tableau. These tools are easy to use and support a large collection of visuals. These tools can be differentiated based on several parameters, including infrastructure support, dashboards, data sources, visualizations, customer/technical support, and pricing.

Some of the key features of Power BI include free sign-up, the ability to import or view data from multiple data sources, the ability to obtain key metrics of business, Quick insights, and data-driven decision making from anywhere. In addition to the key features, it also supports several advanced features including the ability to embed Power BI reports and dashboards into a Web App, real-time streaming, support for natural language query, share content pack, and integration with Cortana.

There are two variants of Power BI including Power BI Desktop and Power BI Service. Power BI Desktop is an on-premises version of Power BI, which allows users to build reports, queries, and data connections. Power BI Desktop can be installed on a local machine. You can connect to any data source and import the required tables from the selected data source to Power BI Desktop. After importing

DOI 10.1515/9781547400720-006

data into Power BI, you can create visuals based on that data. Once the required visuals are added to the report, you can then save your report.

Power BI Service/Power BI Online is a business intelligence service that hosts reports in the cloud (Microsoft Azure). The major difference between Power BI Desktop and Power BI Service is that the former focuses on creating data while the latter focuses on sharing data.

The major building blocks of Power BI Service are dashboards, reports, workbooks, and datasets.

A dashboard is a collection of tiles that can contain no tiles to many tiles. A Power BI report is a collection of visualizations/visuals, such as charts and graphs. A dataset is defined as a set of data that we import or connect to in Power BI. A special type of dataset that can be imported or connected to Power BI is a workbook.

Reports created in Power BI Desktop can be published to Power BI Service where it can be accessed by other users in your organization.

Power BI Azure Application

Power BI can be integrated with Azure services to generate real-time insights into your business. Azure APIs can be used to view the real-time business data in an easy to understand and highly visualized manner. A Web app created on Azure can be embedded with Power BI to visualize the report created in Power BI. The real-time streaming feature of Power BI allows business analysts to collect real-time information from different sources that provide time-sensitive data. For visualizing real-time data, you are required to set up the real-time streaming dataset in Power BI. Power BI supports three datasets including Push dataset, Streaming dataset, and PubNub streaming dataset. You can push data into a dataset using Power BI REST APIs, streaming dataset UI, and Azure Stream Analytics.

The Quick Insights feature of Power BI applies complex algorithms to the dataset and locates different subsets of the dataset quickly within the specified time frame.

Power BI on Microsoft Stack

Power BI can be integrated with a large number of data sources including Excel, SQL Server, PostgreSQL, Dynamics CRM, MySQL, etc. One data source provided by Microsoft is Microsoft SQL Server. Microsoft provides a large list of features with this integrated solution because both are products of Microsoft. You can use both the Import and the DirectQuery options to import data into Power BI from SQL Server. Once you import data into Power BI, you can establish a relationship between tables.

You can also use Data Analysis Expressions (DAX) expressions to create calculated columns and tables. DAX is an expression (a set of functions, constants, and operators) written to apply calculations on data available in your model. Once the calculated tables and columns are ready, you can use the data to create visuals for the report. You can easily publish this report to Power BI Service. To access the report from anywhere, you can set up a gateway on the machine running SQL Server. A gateway is like a bridge that establishes a connection between Power BI and SQL Server. It is a piece of software that allows users to access data located on an on-premises system or network so that it can be used in a cloud service later. You can configure the data gateway for Power BI and add a data source to it. You can also use the data refresh feature in Power BI. You can set a scheduled refresh of data so that your Power BI report visualizes updated information. For successful configuration of a scheduled refresh, you need to set up the gateway connection, data source credentials, and schedule refresh.

A content pack is a complete package of your dashboard, report, and dataset that can be shared with other users in your organization. You can create the content pack and publish it to the team. You should have a Power BI Pro account for creating and accessing an organizational content pack.

Power BI can also be integrated with one of the most advanced features of Windows 10, that is, Cortana Intelligence Suite. When you integrate Cortana with Power BI, Cortana also looks into Power BI dashboards and reports for related keywords each time you make a query to Cortana.

Power BI on Open Source Stack

PostgreSQL is an open source object-relational database management system (ORDBMS). Power BI can be integrated with PostgreSQL and thus you can create intuitive reports. To integrate PostgreSQL with Power BI, you should have PostgreSQL, PostgreSQL database, Npgsql connector, and Power BI. Npgsql connector is an open source ADO.NET Data Provider or connector that allows Power BI users to connect to PostgreSQL database.

To get data from PostgreSQL to Power BI, you first need to install the Npgsql connector. Once the data is available in Power BI, you can apply data modeling strategies to the data. After the data modeling is done, you can create intuitive reports based on the filtered data. You can save it and then publish it to Power BI.

You can set the visuals to display updated data by applying manual refresh and scheduled refresh. To apply scheduled refresh, you need to set up a gateway connection. As mentioned earlier, a gateway is a piece of software that allows users to access data located on an on-premises system or network so that it can be used in a cloud service later.

You can create a content pack that is a complete package of your dashboard, report, and dataset and can be shared with other users in your organization. As you publish the content pack, it becomes available in a centralized repository called AppSource.

Power BI on ERP Applications

One of the most prominent and powerful features of Power BI is its integration with Dynamics CRM. The ability of Microsoft Dynamics CRM to enhance customer relationships for any organization make it a perfect choice for entrepreneurs to choose it over other data sources.

Customer relationship management (CRM) manages interactions between a company and its customers. A CRM solution simplifies processes and enhances profitability in different divisions including sales, marketing, and service divisions. It also centralizes customer information and automates marketing connections.

Microsoft Dynamics CRM improves customer relationships for any type of organization. It is based on the extended relationship management (xRM) plat-form and allows partners to customize it through the .NET based framework. It is easy to use due to its simple user interface (UI).

Microsoft allows organizations to use Microsoft Dynamics CRM solution on a trial basis and then requires a company to use paid services as per business requirements. You can easily get data into Power BI using the Get Data button. Once the data is in Power BI, you can create reports.

The Deep Linking feature allows users to view the information available on the Dynamics CRM portal by clicking the available link.

Row-level security (RLS) is a feature of Power BI that filters the content based on a user's role. You can create roles in Power BI Desktop. These role definitions are automatically published to Power BI Service when you publish your report to Power BI Service.

After creating a role in Power BI, you need to add members to the specified role to allow them to see the filtered results specified in RLS. Once you are done with all your work on the report, you can easily share it with your colleagues.

Index

DOI 10.1515/9781547400720-007